D0098500

# STORIES IN STONE

BY THE SAME AUTHOR

*The Seattle Street-Smart Naturalist*

*A Naturalist's Guide to Canyon Country*

# STORIES IN STONE

*Travels Through Urban Geology*

DAVID B. WILLIAMS

WALKER & COMPANY
NEW YORK

Copyright © 2009 by David B. Williams

All rights reserved. No part of this book may be used or reproduced in any manner whatsoever without written permission from the publisher except in the case of brief quotations embodied in critical articles or reviews. For information address Walker & Company, 175 Fifth Avenue, New York, New York 10010.

Published by Walker Publishing Company, Inc., New York

All papers used by Walker & Company are natural, recyclable products made from wood grown in well-managed forests. The manufacturing processes conform to the environmental regulations of the country of origin.

LIBRARY OF CONGRESS CATALOGING-IN-PUBLICATION DATA.

Williams, David B.
Stories in stone : travels through urban geology / David B. Williams.
p. cm.
Includes bibliographical references and index.
ISBN 978-0-8027-1622-4 (hardcover)
1. Urban geology—United States. 2. Urban geology—Italy. I. Title.
QE39.5.U7W55 2009
550.9173'2—dc22
2009005609

"To the Rock that will be a Cornerstone of the House" from *The Collected Poetry of Robinson Jeffers: Volume One, 1920–1928*, edited by Tim Huntby. Copyright 1938, renewed 1966 by Donnan and Garth Jeffers. All rights reserved. Used with the permission of Stanford University Press, www.sup.org.

All photographs by David B. Williams except the following: Page 13, photograph by David B. Williams, used by permission of the Amherst College Museum of Natural History/The Trustees of Amherst College. Pages 29 and 193, photographs by Adam Shyevitch. Page 41, photograph courtesy of the Thomas Crane Library, Quincy, Massachusetts, Parker Collection. Page 65, photograph by Horace Lyon, used courtesy of Archival Collection, Tor House Foundation. Page 93, photograph by John Cipriani, used by permission of National Park Service.

Visit Walker & Company's Web site at www.walkerbooks.com

First U.S. edition 2009

1 3 5 7 9 10 8 6 4 2

Typeset by Westchester Book Group
Printed in the United States of America by Quebecor World Fairfield

To Marjorie

CONTENTS

# Preface

Most people do not think about geology when they are walking on the sidewalks of a major city. But wherever I am in the world, whether strolling through downtown Boston or hiking in the North Cascades, rocks are the first thing I see. I can't walk by a beautiful stone building, or even an ugly one, without touching it and trying to figure out where the stones came from. I go to cemeteries because of the unusual rocks used as grave markers. When I travel I bring back rocks for souvenirs, and I ask friends to pick up rocks from their exotic vacations. When I watch movies, my eyes wander to the buildings in the background to see what stone I can identify.

My passion for rocks began when I majored in geology in college. I had originally planned to go into engineering, but when I took physics and got 16 percent on a quiz I turned to a subject I had excelled at: field trips. I loved being outside learning about the planet and its history. After college, I moved to southern Utah. My passion for stone grew as I hiked, biked, and explored the red rock landscape around Moab, a geologic paradise.

When my wife, Marjorie, decided to pursue her master's degree, we moved to Boston. I hated the first few months. Where I had once traipsed through quiet sandstone canyons, surrounded by thousand-foot-tall cliffs of rock, I now walked through shadowy canyons created by buildings. Where I once hiked on desolate trails, I now crossed busy streets. For the first time in many years I felt disconnected from the natural world.

And then I noticed Boston's buildings. Half-billion-year-old slates abutted 150,000-year-old travertines. Sandstone that formed in Connecticut sat on top of marble that formed in Italy. Metamorphic rocks interfingered with igneous rocks. Fossil-rich, sea-deposited limestones juxtaposed mineral-rich, subduction-created granites. Plus, builders had gone to the effort of cleaning and polishing these fine geologic specimens, making their stories that much easier to read. As I began to notice the stone in buildings, I found the geologic stories that could provide the connection to wildness I had lost.

Now when I need a rock fix, I simply go downtown and wander the business district and look at building stones. I see evidence of continents splitting apart, crashing into each other, and diving deep into the planet; of rivers washing into dinosaur-rich valleys, seas teeming with invertebrates, and hot springs bubbling with bacterial stews; of magma baking limestone into marble, granite carried thousands of miles by plate tectonic movement, and massive trees fossilizing into stunning pieces of petrified wood.

Geology is half the story; building stones also provide the foundation for constructing stories about cultural history. Granite used at the Bunker Hill Monument led to the construction of America's first railroad. In the brownstones of New York and Boston, the whims of fashion combined with the realities of weathering and erosion to dictate over three hundred years of architectural planning. And in the cottage and tower poet Robinson Jeffers built for his wife and family, you can see how his years of intimate work with stone grounded Jeffers and helped make him one of the most popular American poets of the early twentieth century.

Whether it is a lone iconoclast building his home with rocks he found on a beach or a multinational conglomerate importing millions of tons of marble, we employ building stone to convey sentiments as mundane or grand as we desire them to be. Our use of stone reflects the long relationship between people and nature—on levels scientific, emotional, and philosophical. It is a relationship that continues to evolve and to inform our actions.

*Stories in Stone* offers readers a new window through which to look at urban landscapes. Intriguing cultural and natural history stories are no farther than the nearest building. Building stone did not change the world. Building stone *is* the world. And it will outlast us all.

# 1

# "The Most Hideous Stone Ever Quarried"—New York Brownstone

*In some thirty years every noble cliff will be a pier, and the*
*whole island will be densely desecrated by buildings of brick,*
*with portentous facades of brown-stone, or brown-stonn,*
*as the Gothamites have it.* —Edgar Allan Poe,
*Doings of Gotham,* May 14, 1844

*Little I ask; my wants are few;*
*I only wish a hut of stone,*
*(A very plain brown stone will do,)*
*That I may call my own;—*
*And close at hand is such a one,*
*In yonder street that fronts the sun.*
—Oliver Wendell Holmes, "Contentment," 1858

IF AMERICA IN the late 1800s was the land where the streets were paved with gold, in Brooklyn and Manhattan the streets were lined with chocolate: chocolate stores, churches, and mansions. But above all, row houses, better known as brownstones. They stretched for block upon block, mile after mile, wearying the eye, as one detractor sneered.[1] In 1880 78 percent of the stone structures in Manhattan had a front of brownstone; in Brooklyn, 96 percent. Brownstones housed the poor, the growing middle class, and the titans of the Gilded Age, such as William Astor II, J. P. Morgan, and Jay Gould.

The brown sandstone (hence brown-stone and later brownstone) that faced most of the row houses gave the architecture its name. The stone served no structural purpose; it was the aluminum siding of the day, a four-inch-thick curtain, attached by mortar to the brick load-bearing walls that supported the wood-frame building. Originally a substitute for more expensive marble and first used for decorative work such as lintels, steps, and quoins, brownstone became the most popular building stone in New York by the 1850s and so synonymous with row houses that any row house, whether clad in brick, limestone, or marble, was called a brownstone.

Typically built on a one-hundred-foot-deep lot, brownstone row houses ranged from twelve to twenty-five feet wide and were three to six stories high. A below-grade servants' entrance led to the basement, where the domestics cooked and the family ate. Homeowners, or in many cases tenants, entered via a steeply staired stoop, an architectural feature brought by the original Dutch settlers who had required a rise to get above their flat homeland's perennial flooding. In America the stoop, née *stoep*, helped people rise above street odors generated by the horse-powered transport system. Later the stoop became a favored place for sitting, hanging out, gossiping, and that venerable Brooklyn pastime, stoopball.

The stoop led to the building's formal entrance, often the finest architectural feature of a brownstone. Many entrances had arched openings, a prominent keystone, and an elaborate fanlight. Others were squared off with a detailed console or hood molding. Columns bracketed many doorways and could be Ionic, Doric, Corinthian, or strange combinations thereof. Tall, arched double doors opened into a vestibule or long hallway and led to the first-floor parlor. Oddly, the parlor may have been the least-used part of the home. Society dictated an entertainment space, but most middle-class people rarely used it. Bedrooms and sitting rooms were on the upper floors. Servants lived in the garret. A cornice, often of iron pounded, sanded, and painted to resemble stone, topped the structure.

Long and narrow, particularly in later years when land prices forced developers to squeeze houses to nearly claustrophobic widths, brownstones would not win a modern-day design competition. Windows illuminated only the front and back rooms; middle rooms were gloomy caverns. Long flights of steep, poorly lit stairs connected the floors, providing a practical way to meet one's daily exercise needs, although one nineteenth-century critic wrote that this progenitor of the Stairmaster led to "fruitless and health-destroying labor."[2]

And yet, to have a brownstone-fronted home was to live the good life in Manhattan and Brooklyn, as well as Boston, Hartford, and Philadelphia. Whether rich or poor, immigrant or Knickerbocker, every level of society aspired to live in a brownstone, called by one writer an "almost proverbial synonym for all that is elegant and desirable."[3] So elegant and desirable was brownstone that the world's richest man, William H. Vanderbilt, who inherited ninety million dollars in 1877, disregarded his architects preferred construction material, white limestone, and built between 1879 and 1882 a pair of brownstone mansions on Fifth Avenue. Designed by the firm Trench and Snook, the twin estates cost two million dollars and reportedly required an around-the-clock crew of six hundred to seven hundred workers to build.[4]

Despite Vanderbilt's multimillion-dollar stamp of approval, brownstone did not fare well with critics. They thought the stone cold and unattractive. Brownstone was only a veneer, they sneered, a pretense. Plus, it didn't carve well, didn't age well, and made buildings look bloated. In the words of Edith Wharton, the city was "cursed with its universal chocolate-coloured coating of the most hideous stone ever quarried."[5]

The critics were partially correct. Water and ice can penetrate and weaken the stone's sandy layers, which peel off the building like sunburned skin. But the stone does not deserve all of the blame: Brownstone failed because builders used poor quality stone or laid it incorrectly. Properly placed brownstone blocks in 150-year-old buildings show little or no degradation.

Its reputation weakened by the critics' complaints, brownstone suffered a decisive blow to its popularity at the Chicago World's Fair of 1893, commonly called the White City. Daniel Burnham's array of snow white, classical buildings returned marble and limestone to their former place of supremacy in the building world. White was in and brown was out. Brownstone would not become popular again for another hundred years.

I first fell in love with brownstone in 1996. Marjorie and I had recently moved to Boston so she could attend graduate school. During my regular walks to get away from our dreary apartment, I wandered among town houses made of brownstone in the Brahmin bastion of Beacon

Hill. The brownstones were distinctive, exhibiting warmth and character. I liked that I could see geologic features such as bedding and erosion, and during those occasional sunsets when the light bounced off low clouds, the stones seemed to be lit from within.

Several months after moving to Boston, I chanced upon what would become one of my favorite brownstone buildings. Built in 1766, Harvard Hall sits on the western edge of the Harvard campus. It is a stately Georgian structure with simple lines, arched windows, and a two-story portico. Although primarily brick, Harvard Hall sits on a base of brownstone, with brownstone stairs that gently rise on either side of the main door.

I distinctly remember walking up to the stairs to look at the stone work, which had been laid incorrectly and had begun to succumb to weathering. Making sure that no one was looking, I stroked the crumbling stone. Sand grains accumulated in my hand. They immediately transported me back to my beloved Utah.

Although I had looked at brownstones for months, it wasn't until the sand grains of Harvard Hall nested in my hand that I made the connection: What I had known as red rock in Utah, easterners called brownstone. Both are sandstone colored by iron, which in an oxygen-rich environment rusts and coats individual sand grains like the skin of an apple.[6] I learned later that my favorite Utah rock layer, the Wingate Sandstone, and Harvard Hall's brownstone were deposited at the same time, during the waning days of Pangaea, and that both contain dinosaur tracks, possibly from the same species. I finally felt like I was making a connection to Boston.

A decade later I headed to the heart of brownstone country, New York City. Specifically, I went to Brooklyn, because many of the brownstone row houses and most of the great brownstone mansions in Manhattan were gone. Some neighborhoods, such as Greenwich Village, the Upper West Side, and Harlem, still have chocolate row houses, but these isolated pockets cannot compare with the street upon street of brownstone-fronted buildings found throughout Brooklyn.

I stayed in a Brooklyn brownstone with a childhood pal and her family. It was a classic five-story row house, with one apartment on each floor. My friends lived on the top floor, so I got to experience the fruitless and health-destroying labor of stair climbing.

The morning of my first day in Brooklyn, my friend Megan and I strolled through the Clinton Hill, Fort Greene, and Brooklyn Heights neighborhoods. Walking through canyon after canyon of sandstone felt like returning to Utah, except I saw a lot more people and I could buy good pizza. Some buildings had smooth, precisely cut rock; others had rusticated blocks; and a few had stippled stones. Moss and lichen covered many of the low walls that fronted patio spaces next to stoops. A few walls had a black patina similar to the desert varnish that coats Utah's cliffs.

A fractured brownstone pile next to a stoop vividly reminded me of a cliff of rock slowly decaying and crumbling. I grabbed one piece and put it in my backpack. I took dozens of photos of buildings where wind and water had smoothed elaborate window and door details to shapeless blobs. On one building, erosion had so weakened the rock that the front door console had fallen off, revealing the brick and wood structure underneath.

Across the street from Fort Greene Park, on South Portland Avenue, was a classic stretch of row houses, their repetitious and receding lines of stoop, doorway, and window looking like an art student's first attempt at perspective drawing. Despite the uniformity of the buildings, they formed a balanced and elegant beauty, further enhanced by the block's tall London plane trees and bluestone sidewalks.

One of the chief complaints of nineteenth-century detractors was the tedium of brownstones. Junius Henri Browne, a critic wearied by brownstone, wrote in 1869, "One longs [for] . . . some change in the style and aspects of the sombre-seeming houses, whose occupants, one fancies from the exterior, look, think, dress and act alike."[7] Perhaps in Browne's day and in the subsequent decades, regularity doomed the buildings, but now this same consistency stands out as a coveted feature. *Time Out New York* recently named South Portland Avenue the best block in New York, citing its unbroken row of brownstone town houses.

South Portland was a wonderful block, but it also showcased a disturbing trend. Many of the buildings no longer had brownstone facing. Instead, stucco covered them. This process of putting stucco over stone began in the 1960s.[8] In neighborhoods where brownstones were run down and unsafe, a few pioneers recognized the beauty and history of the buildings, which were cheap and available even if bankers wouldn't extend credit. Buyers had to move fast because city agencies had started to demolish brownstones. Ironically, demolition led to the open spaces now utilized as community gardens throughout Brooklyn.

When the new residents began to move into these neighborhoods, they began to repair and restore the buildings. "People didn't have the

South Portland Avenue in Brooklyn.

money they have now, so there was less attention to detail. In many cases, owners simply removed or shaved off features, in particular around windows," said Alex Barrett, an architect who focuses on brownstone restoration in Brooklyn.[9] Where once a prominent sill and ornate molding and brackets surrounded a window with distinction, plain windows now punctuated buildings with flaccid monotony. Owners also lopped off stoops, abandoning the original entryways like forgotten lovers.

To stave off erosion of the stone, brownstoners, as the new immigrants dubbed and still dub themselves, also painted brownstone facades. In a second mistake from the 1960s, many brownstoners used nonbreathable paint, which trapped moisture and exacerbated stone spalling. Other brownstoners patched the weakened stone with stucco. Often, however, they failed to match the colors of stucco and stone, leaving the building looking like a teenager's blotchy face. With more experience and money, later owners began to stucco the entire building, and it is these more modern restorations that rankle brownstone purists.

"We recently restored a five-story building and spent seventy-five thousand dollars, which was a good price. We could have paid much less or much, much more. It took two to four guys working full time four months to complete it," said Barrett. First, restorers jackhammered off the old stucco to get down to the original stone. They then applied a gray cement mortar mixed with coarse aggregate to form a base, or scratch coat. Barrett's restorers hand molded all of the scroll work around the new windows using simple trowels. They rebuilt the steps by hand with stucco. On the finish stucco coat, they applied an aggregate-free cement mortar with a custom-created mix of colored sand.

The artistry of a high-quality restoration such as the one on Barrett's row house bestows a dignity that harkens back to the glory days of the brownstone, when owning one meant that a person had achieved a certain level in society. The fancy detailing around the windows and doors gives the building elegance, style, and depth. Barrett's attention to detail results from his interests and concerns about history and the importance of brownstone to the development of New York. But I miss the imperfections of true brownstone. Most restorers, at least the high-end ones, do such a good job that the buildings lack character. The lines are too straight, the stucco too homogenous, the color too even. While stucco restoration corrects the fatal flaws of the past, the buildings lose their soul. The great row houses of modern Brooklyn are no longer brownstones but "brownstuccos." They could fit right in in Santa Fe, stucco capital of the world.

What I like best about brownstone is its geologic essence. When you look closely, you can see that the individual sand grains vary in size from mote to pebble, and in color from reddish to deep mocha. Some bedding planes are thick, some are wavy, and some are not visible because the builder placed the stone with the bedding plane face out, which tends to make the blocks look like wood grain. The erosion differs, depending on resistance and aspect. I found one building with a pair of dragon faces carved out of brownstone below the front porch. One retained its detail while the other had worn away to a ghost of its original fierceness. This heterogeneity reflects the original, complex depositional environment of the stone 200 million years ago.

Most people do not encounter geologic phenomena on a daily basis. You may read about distant volcanoes, earthquakes, and tsunamis in the news but with brownstone you can see the same processes that wore down the Appalachian Mountains and carved the Grand Canyon. You can see how water and ice infiltrate and ferret out the weakest links in a rock and slowly reduce it to its constituent grains. A solid in geologic time is not truly a solid, and it will surrender to an overriding principle of nature—gravity; what goes up must come down, even if it takes millions of years or in the case of the hapless brownstones, decades.

The basic geologic story of brownstone is simple and appealing. Go back 200 million years. Streams wash into a valley and deposit layer upon layer of sand and silt. Dinosaurs plod through the wet sediments leaving behind thousands of tracks. Sediments and tracks harden into sandstone. To understand why dinosaurs inhabited that valley, why quarries occur where they do, and why brownstone was a good building stone, however, requires adding a few more details.

The valley where the dinosaurs roamed sat in the middle of the supercontinent of Pangaea, which like a giant puzzle, consisted of many smaller pieces of land. The northern portion, called Laurasia, included Siberia, Europe, and North America. Antarctica, Africa, South America, India, and Australia, collectively known as Gondwana, formed the south part of Pangaea and extended down to the South Pole. All was not right, however. As that great geologist Bob Dylan sang, "He not busy being born is busy dying," and Pangaea began to break up.

At least fifteen rift valleys, or basins, opened as North America, Africa, and South America pulled away from one another. The individual basins stretched from Alabama to the Bay of Fundy. Streams from adjacent hills and mountains began to carry sand and silt into the basins, including one where the Connecticut River now flows, from about modern-day Amherst, Massachusetts, to New Haven, Connecticut. Geologists call this lowland either the Connecticut River valley or the Hartford Basin.

Between about 220 and 195 million years ago, this valley lay about ten to fifteen degrees north of the equator, roughly the same latitude as present-day El Salvador. Dry and warm with an ecosystem of ferns, cycads, and conifers, the lowland received less than twenty inches of rain per year, mostly as seasonal monsoons. A variety of small dinosaurs, about six feet tall and shorter, ten-foot-long amphibians, and fish-eating crocodilelike animals inhabited the valley.

And then, as the continents continued to pull apart, Earth's crust thinned and the Hartford Basin ripped open, like an overstuffed sausage.

Black lava spread from swarms of fissures in Connecticut and all of the rift valleys that stretched for a thousand miles along the eastern margin of North America. With a consistency of ketchup, the basalt flowed thousands of yards per day. In addition to wreaking havoc on the landscape, the viscous basalt spewed out trillions of tons of sulfur dioxide and carbon dioxide, generally making the planet an unpleasant place for any species that liked clean air.

Geologists speculate that this worldwide flood of basalts may have contributed to a mass extinction of 50 percent of planetary life, including a diverse group of carnivores and herbivores, generally bigger and badder than dinosaurs who lived at the time. With their competitors out of the way, dinosaurs, which had first evolved about 30 million years earlier, reacted quickly and doubled in size. They also began to evolve into the myriad species that dominated Earth for the next 140 million years. Within twenty-five thousand years of the extinction, new forms had emerged including *Anchisaurus*, a long-necked herbivore, and twenty-foot-long predators such as the double-crested *Dilophosaurus*, one of the stars of *Jurassic Park*.[10] They had taken over from slim, three-foot-long plant eaters and similarly sized meat eaters. Nowhere on the East Coast is this record of dinosaur ascendancy better recorded than in the fifteen thousand feet of sediments that accumulated in the Connecticut River valley.

Of all the rock that formed in the valley, the thickest, youngest, and most important to the brownstone story is the Portland Formation. Named for its main point of origin—the town of Portland, Connecticut—it is the rock unit that provided the building blocks for most of the brownstone row houses of New York and Boston. The stone formed very rapidly, in just a few million years, as streams carried sediments out of the surrounding highlands and into a valley of lakes, floodplains, and river channels.

The warm and well-watered valley was ideal habitat for dinosaurs. As they tromped around in the moist mud and sand along the valley's streams and lakes, the great and the small left behind thousands and thousands of footprints, which remained intact as the mud hardened to rock. These tracks are one of the coolest and also most geologically important aspects of the brownstone. Because fossilized footprints record a specific moment in time in the life of an animal, they eventually helped paleontologists revolutionize our understanding of dinosaurs.

★ ★ ★

In 1802 a lad named Pliny Moody was working his family's field in South Hadley, Massachusetts, when his plow thumped against a block of brown sandstone. Clearing away the soil, Moody discovered four raised tracks crossing the flat slab. Each four-inch-long track had three toes and looked like it was made by a bird. Since the slab was of no use in the field, Moody's family decided to use it as a doorstep, where it remained for several years until a local doctor, Elihu Dwight, purchased the curious rock. Dwight nicknamed the tracks' maker Noah's Raven, and showed them to Amherst College natural history professor Edward Hitchcock, who also thought that birds had made the tracks.[11]

The Noah's Raven slab is now on display in an honored spot in the main collection of tracks at the Amherst College Museum of Natural History. Three feet by two feet by two inches thick, the reddish slab tapers to twelve inches wide at the bottom, where the top two inches of a toe are visible. Three other tracks run in a line up the slab, clearly showing where a dinosaur walked across the wet sand. The tracks are darker and raised slightly above the surrounding rock. They aren't actually an impression but a positive cast of the original track. Dinosaur tracks form when an animal steps in moist, firm sediment, which subsequently dries and hardens. New sediment then fills in the track, creating a cast, as well as preserving the original. Geologists generally refer to both the cast and the original as tracks.

Hitchcock, who acquired the slab from Dwight, was close to correct about what walked around in the mud of the ancient valley. A dinosaur, the progenitor of birds, made the Noah's Raven tracks about 200 million years ago. It stood about thirty-six inches tall and walked on two feet in a pigeon-toed manner. Other tracks from this dinosaur species indicate that they could have walked on all fours, occasionally dragged their tails, rested with their breast and rump on the ground, and traveled in family groups. They also fidgeted, or at least multiple prints in the same locality indicate that they stomped or patted their feet when resting. The same type of tracks have been found across eastern North America, in South Africa and Poland, and on the Colorado Plateau. Paleontologists call the track maker *Anomoepus*, meaning "unlike foot."[12]

Noah's Raven is only one part of Amherst's collection of twenty thousand tracks and casts. Most of them came from the Connecticut River valley, either from the Portland Formation, including the Noah's Raven slab, or equivalent rocks in other parts of the valley in Massachusetts. The collection, the world's largest, was assembled by Edward Hitchcock between 1836 and 1864 and now resides in a beautiful new museum on the Amherst campus.

"Hitchcock was the preeminent geologist in America," said Steve Sauter, coordinator of education at the Amherst museum.[13] And yet Hitchcock never conceded that the Noah's Raven tracks were made by a dinosaur and therefore were the first evidence of dinosaurs found in America. (British naturalist Richard Owen coined the term dinosaur in 1841 and by the 1850s most geologists and naturalists knew dinosaurs to be a widespread and diverse group.) "Hitchcock could never accept that God created monstrous beasts like dinosaurs," said Sauter. "He always thought that birds made the tracks."

Hitchcock was a bizarre mixture of scientist, puritan, hypochondriac, and country bumpkin. Born in 1793 in Deerfield, Massachusetts, he only attended six winter terms at Deerfield Academy, which at the time was a K–12 public school. He didn't go to Harvard or Yale because he got it into his head that he was too sickly to attend. Other people in town helped look after his intellectual interests, however, lending him books on Latin and Greek and talking to him about subjects such as military tactics. He became a Congregational minister but with minimal training. He didn't drink alcohol, ate no meat, and subsisted on a cornmeal gruel made with tepid water.

Hitchcock had an intense interest in science. He started to conduct his own experiments and reported his findings to professional scientists such as Benjamin Silliman, Yale's influential professor of chemistry, mineralogy, and geology. Word got out about the scientifically inclined minister, and in 1825 Amherst offered him a job teaching theology and science. Hitchcock knew he knew theology but was unsure about science, so he appealed to Silliman for help. He spent several months learning to teach science from Silliman and returned to become Amherst's professor of chemistry and natural history. He remained at Amherst until his death in 1864.

Ten years after starting at Amherst, Hitchcock's life changed when he received a letter from a local doctor, James Deane. Deane's letter described "the tracks of turkeys in relief" from a slab of sandstone about to be used as a sidewalk in Greenfield, Massachusetts.[14] Hitchcock ignored the letter until Deane finally sent plaster casts of the tracks. Now convinced that they were tracks, Hitchcock traveled to Greenfield to see the slab. Within a year he had found more tracks in sidewalks in

Northampton and Deerfield, as well as in several quarries. He had also seen the Noah's Raven slab and collected enough additional tracks to publish America's first scientific paper on fossil tracks. He called the nascent field ornithichnology, the study of stony bird tracks.[15]

Hitchcock's paper described seven species of track makers. The casts and impressions ranged in size from the four-inch-long Noah's Raven to the massive *Ornithichnites giganteus* (gigantic stony bird tracks) with fifteen-inch-long feet and a six-foot-long stride. Tracks of several different species walking in different directions covered one slab. Another showed one animal leaving over ten tracks in a steady line and most were so distinct Hitchcock could determine the left and right foot. He concluded that "they could not have been made by any other known biped, except birds."

Smitten with tracks, Hitchcock started collecting them himself. He always wore his black suit and tie when out in the field, although often he would sneak home late at night because he recognized that digging and transporting tracks was "not comporting with the dignity of a professor."[16] He even found and made a cast of tracks from a sidewalk on Greenwich Street in Manhattan. Hitchcock later wrote that casting the Greenwich tracks almost landed him in the local asylum: A former student saved him when she testified that he was "no more deranged than such men usually are."[17]

His favorite track slab came from the Portland brownstone quarry. It had been used for decades as a sidewalk, with the nontrack side facing up, until yet another local doctor heard of Hitchcock's interest in tracks and remembered he had seen unusual markings on the slab when it had been laid in place. (I wonder what all the sick people were doing while these doctors were searching for tracks.) Called by Hitchcock the "gem of the Cabinet," it shows mud cracks, worm tracings, and 54 beautifully preserved track casts of several species.[18] By the time he died, Hitchcock had named 216 species from thirty-eight localities and published more than thirty reports, including his magnum opus, *Ichnology of New England*, which contained some of the first photographs taken of fossils.

Hitchcock's work took place at a critical time, when geologists were starting to refute the accepted dogma of the biblical stories of Adam and Eve, Noah's flood, and God's creation. In 1815 William Smith published the world's first geological map, which showed the geology of England and Wales. Smith based his work on his observation that sedimentary strata contain fossils that occur in a definite, predictable sequence and

Edward Hitchcock's "gem of the Cabinet," catalog number 9/14.

that these layers could be correlated between locations. Combined with Nicolas Steno's law of superposition, which states that older rocks lay under younger rocks, Smith's work made geology into a three-dimensional science based on descriptive analysis instead of pure speculation.

Fifteen years later, Charles Lyell published his seminal work, *Principles of Geology*, which argued that natural laws did not change over time, therefore modern geologic processes acted in the same way and at the same rate as they had in the past. Lyell's book helped establish that Earth was not created six thousand years ago but must be very old because geologic phenomena, such as erosion and deposition, occurred so slowly that vast expanses of time were necessary to produce the planet's varied landscapes.

A third great advance came from Swiss-born geologist Louis Agassiz, whose *Étude sur les glaciers* in 1840 established the importance of ice in sculpting landscape. Agassiz showed that a great and geologically recent ice age was responsible for ice sheets that carved valleys, shoved moraines, and carried erratic boulders. His work was another critical step in helping to dispel the myth that catastrophes (i.e., biblical deluges) were responsible for modern geologic features.

Finally, with Charles Darwin's *On the Origin of Species*, published in 1859, geology also started to address the great biological questions. Through careful observation, accumulation of data, and formation of testable theories, geologists of the nineteenth century opened major doors in understanding the history of Earth. Plants and animals evolved and went extinct. Landscapes changed, sometimes drastically, over time. Earth was a very old planet. These are the central themes that still drive geology.

"It's just stunning that all of this is coming together. That the science of geology was just exploding," said Sauter. "Western travelers are bringing back all of these fossil specimens and animals, like dodo skeletons and moa skeletons in the Pacific. It's all coinciding and all clashing with this biblical belief and there's Hitchcock in the middle of this storm. And he is the first person to have the imagination to question the tracks. To ask, What kind of animals made these footprints? How could prints be made in stone? How old are these footprints? He essentially creates an entire new field of science."

When Hitchcock died, however, interest in the tracks faded. A year after his death, the Civil War ended, people began to move west, and they discovered hordes of dinosaur fossils. Hitchcock's tracks could not compete with the bones of *Tyrannosaurus rex*, *Stegosaurus*, and *Triceratops*. But in the 1970s and 1980s, paleontologists returned to the tracks. They began to ask new questions about dinosaur behavior.

Unlike bones, which tell the story of death, tracks record the action of a living animal. Tracks show young and old dinosaurs of the same species traveling together, different species visiting the same shoreline on the same day, and dinosaurs following each other. "In the 1950s we thought that dinosaurs were sluggish, solitary creatures that dragged their tails around behind them," said Sauter. "And now we think of them as athletic and birdlike. They were particularly vicious, and fast runners and jumpers. We found out all this information from these slabs. These actual slabs. It was really a revolution of thinking."

"All the right things came together at the right time for brownstone," said Alison Guinness, whose interest in the rock began when her master's thesis adviser at Wesleyan University received a grant to study the Portland quarry. "It was easy to transport. There were a large number of workers available. Demand was growing, the rock was easy to quarry, and there was a lot of it."[19]

During three centuries of quarrying in Portland, workers extracted more than 270 million cubic feet of rock, or enough material to build three copies of the Great Pyramid of Giza.[20] It was shipped up and down the East Coast, overland to Chicago, and around Cape Horn to San Francisco. "Brownstone was so important to the local economy that they even moved a cemetery to get at the rock," Guinness said, standing next to the old quarry in Portland.

She knows the history of these quarries better than anyone. When she began studying them, they had been forgotten and abandoned for years. "They were languishing. They had filled with water and people dumped cars in them," said Guinness. "I quickly learned, however, that these quarries were the ultimate site that shaped the entire brownstone industry."

In 1686 James Stancliff became the first person to settle on the east side of the Connecticut River in what is now Portland. The selectmen of Middletown, located on the west shore of the river, had given him the land called The Rocks so he could harvest stone to build chimneys and cut gravestones. Stancliff's sons also joined the business, which they later sold to another family of gravestone cutters, led by Thomas Johnson. Their tombstones were the first Portland stone to be exported widely and show up in cemeteries as far away as Newport, Rhode Island.

Guinness noted that Stancliff was not the first person to recognize the importance of the rocks that outcropped next to the Connecticut River. After Middletown's settlement in 1650, locals had used the stone for foundations, steps, and walls. Town residents probably didn't quarry the rock but simply pried the stone from ledges along the water and carried it away in carts and scows, said Guinness. As so often happens, word got out and non-Middletownians began to arrive at The Rocks with their own picks, carts, and watercraft to remove stone.

Responding to such effrontery, the fine citizens of Middletown voted on September 4, 1665, "that whosoever shall dig or raise stone at ye rocks on the East side of the Great River . . . the diggers shall be none but an inhabitant of Middletown and shall be responsible to ye town twelve pence per tunn . . . to be paid in wheat and pease . . ."[21]

The high value of grain and green vegetables seems to have quelled the stone stealers. Locals resumed collecting rock and felt so magnanimous with their bounty that they gave Stancliff his one-third acre at The Rocks. Ten years later Stancliff obtained another half acre, but by 1715 Middletownians were worried again. This time they banned even locals from collecting stone and transporting it out of town. Scofflaws had to fork out twenty shillings per stone. Concerned townsfolk also appointed a quarry agent to enforce the rules at what was now known as the Town Quarry.

This round of posturing didn't last long. Using his skills as a gravestone cutter, Thomas Johnson quarried enough stone in 1737 to provide brownstone accents for a granite house in Boston for Thomas Hancock, John's uncle. Other rock began to make its way down to New York and Newport for architectural trimmings, but quarrying stayed small scale because little demand existed and transportation was challenging.

Quarrying in the early 1700s was still a crude affair. If they didn't collect rock from the surface, workers blasted it out with black powder or knocked it off with an ax. They were aided by how the sandstone formed. During the monsoon seasons 200 million years ago, the rivers would flood and overflow their banks, each time depositing a sheet, or bed, of sand as thick as five feet or as thin as half an inch. Bedding created a flat surface ideal for building blocks. Quarrymen also took advantage of bedding because the contact zone between beds is weak—relative to the surrounding rocks—and is easier to pry apart. Furthermore, floods often deposited nearly pure beds of the same-sized sediments, which made the Portland rocks homogenous and easy to work.[22] These beds are clearly visible in the walls of the Portland quarry. The most useful extend for hundreds of yards.

Earlier tectonic change also aided the quarrymen. As Pangaea continued to break up, tectonic forces pushed and pulled the valley. Tension was released by a series of cracks or joints, later known to quarrymen as keys, that tended to run at ninety-degree angles to the horizontal bedding. These seams and bedding created four workable faces. Masons needed only to cut two more faces to form a block. Subsequent generations of quarrymen exploited this geologic warp and weft on a much larger scale.

Commercial quarrying did not begin until the 1780s when several companies began working holes in and near James Stancliff's original Portland site. By the 1820s three companies—Middlesex, Brainerd, and Shaler and Hall—had consolidated ownership of the quarrying business, although Middletown citizens could still take rock from the Town Quarry. The Middlesex and Brainerd pits opened near the public hole and next to each other, separated only by the Portland town cemetery. Shaler and Hall's quarry sat just downriver.

The companies also controlled the town, said Guinness. Workers shopped at the company stores, using credit extended to them by the company. They lived in company housing and many prayed in a church built with stone provided by the companies.

The men worked from sunup to sundown with a midday break of one to two hours. Work stopped for bad weather, holidays, and haying. Guinness noted that "going to the poorhouse, drunkenness, and wife's jollification" also led to missed days, or at least hours, depending upon the activity. Typical wages in the 1830s ranged between $11 and $18 per month. By the 1850s the three companies had conspired to set a standard pay rate, which was $1.10 per day in 1854. Wages peaked at a daily rate of $2.50 in 1870, sliding back to $1.55 in the nationwide depression that followed.

Prior to the use of steam engines, men and beasts did all the work. A full labor force topped out in the 1850s with between twelve hundred and fifteen hundred men, 60 span of horses, and 120 yoke of oxen. Surveyors determined where to quarry. Rock bosses supervised a crew of blasters, cutters, and haulers. Teamsters controlled the oxen and horses that moved the stone within and out of the quarry. Measurers ensured the size of each stone that left the quarry, and blacksmiths tended wagon wheels, picks, drills, and other iron and steel implements. All the while, a diligent timekeeper tracked the comings and goings of the men.

After 1850 the workforce began to shrink, a consequence of quarry owners adopting new technology as soon as it became available, said Guinness. Pumps allowed the quarries to drop below river level by removing excess water. Steam-powered derricks could more easily remove rock from the depths than could animal-powered carts. Derricks also raised and lowered men and oxen. New motorized tools such as jackhammers required fewer men to operate them, while three miles of narrow gauge railroad track and several locomotives made moving stone and cranes less labor intensive. By 1896 only two yoke of oxen remained and about half the number of workers as had been employed in the 1850s.

After the stone had been cut, quarrymen loaded blocks onto shallow-draft schooners called stone boats, or brownstoners, for the thirty-mile trip south on the Connecticut River to Long Island Sound. Towed downriver by steamboat, stone boats reverted to wind power in the sound. A typical trip on the Connecticut River took eight to twelve hours.

Weather controlled the work season, which started after the thaw. Because the quarries spread out next to the river, spring flooding could also dictate when the men worked. For example, on May 4, 1854, the Connecticut River overflowed and completely filled the quarries. It took ten days to pump out the water. The season ended when boats could no longer travel on the iced-over river.

Cold weather also affected the stone. When first quarried, it was saturated with moisture, called sap by quarrymen, which could destroy a rock in freezing weather. During summer and fall, quarrymen seasoned the blocks by covering them with soil and letting them dry for four months. Seasoning case-hardened the stone by allowing dissolved calcite or silica to move with the sap to the surface, where the minerals deposited a new, stronger coating. In later years, during the height of brownstone popularity, demand was so great that quarrymen didn't have time to let the rocks season, which resulted in poor quality stones that helped ruin brownstone's reputation.

And good stone was key. Of the ten million cubic yards of rock removed from the quarries about half was waste, dumped outside the quarry. The quarries annually generated up to two million cubic feet of rock during peak production years, equally divided between high-quality stone, including unseasoned rock, and stone used for nonarchitectural purposes, such as abutments and piers.

Oddly, no stone quarried in Portland was cut and trimmed in town, except for local projects. Most stone was shipped raw to New Jersey or New York for cutting and dressing until 1884 when E. I. Bell established the Connecticut Steam Brown Stone Company, where masons used diamond saws, gang saws, planers, lathes, and a rubbing bed to slice, carve, and finish everything from entablatures to steps to balustrades.

The adjacent Brainerd and Middlesex pits eventually reached down two hundred feet. They became one big hole in the 1870s when the companies purchased the Portland town cemetery, which formed a hundred-foot-high ridge between the quarries. The graves and gravestones—the oldest stone dates from 1712—were moved and now rest a couple miles away at the Episcopal church. All that remains of the ridge is a low, tree-covered peninsula that extends out into the lake that fills the quarry.

Bad weather finally killed Portland's quarrying industry, which had been on the decline since the early 1900s and silent since the 1920s. In 1936 record high water on the Connecticut River flooded the pits. Two years later, a hurricane helped push the river back into the quarries, which have remained flooded ever since. A local company now has the rights to lead diving tours into the quarry lake, which is about 600 yards long by 350 yards wide. Guinness has heard rumors that two train engines might be in the hole, but no one knows all that rests on the bottom of the quarry. In recent years, cleanup crews have removed forty tons of trash, including eight motorcycles, four cars, and sixteen air conditioners, but no trains.

The water-filled quarry now sits a couple hundred feet from the edge of the Connecticut River. Originally the holes were adjacent to the river but quarrymen had simply dumped waste over the western edge of the quarry and created a landfill. Massive baby blue oil tanks and a parking lot guarded by a pair of Rottweilers take up much of the new land. Sumacs, sycamores, and locusts grow on the cliffs above the quarry, their russet, yellow, and red leaves complementing the blue water and brown sandstone.

Few people thought much about the quarries until the mid-1980s, when developers wanted to cut a channel to the river and open a marina

Site of Middlesex and Brainerd quarry,
now flooded, in Portland, Connecticut.

in the lake. They would have been successful except the bottom dropped out of the real estate market and the developers went bankrupt, said Guinness. With her prompting, the city of Portland finally bought the three quarries and adjacent land in 1999 and 2000. They plan on developing the site with trails, educational exhibits, and recreational uses. In April 2000 Secretary of the Interior Bruce Babbitt designated the quarries as a National Historic Landmark.

New Yorkers didn't initially care for brownstone, using it only for building details. For example, the city's oldest church, St. Paul's, built in 1766, is made of 450-million-year-old Manhattan Schist with brownstone quoins, a patio, and columns. The earliest known wall of solid brownstone is part of City Hall (1803–1812), located just north of the financial district. A guidebook published at the time called it "the most prominent and most important building in New-York. It is the handsomest structure in the United States; perhaps of its size, in the world."[23] Hyperbole aside, City Hall is an elegant edifice with a columned entryway, broad stairs, and a slender dome with a skin of white marble from Massachusetts. The builders also used marble on the south, east, and west exterior walls. On the north side, however, the exterior was brownstone, used because no one, or at least no one of any importance, lived north of the building. Those who did either wouldn't know any better or wouldn't mind looking at what builders considered to be a cheap substitute for more classically correct marble and limestone.[24]

As Alain de Botton noted in *The Architecture of Happiness*, for over one thousand years "a beautiful building was synonymous with a Classical building, a structure with a temple front, decorated columns, repeated ratios and a symmetrical façade."[25] Classical architecture began with the Greeks, continued with the Romans, and, following a thousand-year hiatus, reemerged during the Renaissance. The stones of choice for most of the great buildings of antiquity and the Renaissance were marble and a type of limestone known as travertine. How could an architect turn against such a simple equation of beauty?

In New York City in the early to middle 1800s, few architects bucked the tradition. Instead, the change came from laypeople. When Richard Upjohn proposed limestone for Trinity Church in Lower Manhattan, the congregants chose brownstone. Completed in 1846, Trinity Church was the first important building with solid brownstone walls.[26] Charles Lockwood, in *Bricks and Brownstone*, argued that the wealthy, well-read members of the church chose the somber stone because they were at the forefront of the rising Romantic movement, best exemplified by the writings of Andrew Jackson Downing, a nurseryman and one of America's first landscape designers.

Downing's most influential book was *The Architecture of Country Houses*, published in 1850. "No person of taste, who gives the subject the least consideration, is, however, guilty of the mistake of painting or coloring country houses white . . . In buildings, we should copy those [colors] that she [nature] offers chiefly to the eye—such as those of the soil, rocks, wood, and the bark of trees . . ."[27] Downing died in 1852, but his influence continued with his architectural collaborator Calvert Vaux and with Vaux's design partner, Frederick Law Olmsted, codesigner with Vaux of New York's Central Park. The development of gas lighting further enhanced brownstone's reputation because the stone masked the soot produced by gas and coal.

During the 1850s a brownstone fog began to creep across New York. It spread northward through Manhattan as the city grew. It swelled across Brooklyn as the borough became a fashionable suburb. It responded to fashion, changing from Greek Revival to Gothic to Italianate. It responded to money, initially facing row houses and later covering mansions.

Many quarries in addition to Portland opened in response to the growing popularity of brownstone. Most were in the great rift valleys formed during the splitting apart of North America and Africa. Five companies quarried in East Longmeadow, Massachusetts. Hummelstown, Pennsylvania, quarries supplied stone across the state and down to Baltimore. Quarries in northern New Jersey, primarily along the Passaic River in Little Falls, Paterson, and Belleville, provided stone for many institutional buildings in Manhattan, including Trinity Church. All of these quarries produced a rusty red sandstone, sold as brownstone, but none sold as much rock as the quarry in Portland.

The march of brownstone-fronted row houses coincided with a tripling of New York's population between 1840 and 1870, from three hundred thousand to nine hundred thousand. It also overlapped with the growth of an emerging middle class, who wanted to show their prosperity—and a brownstone perfectly served that purpose. For those who couldn't afford an entire building, which included many since the structures were so big, they could rent a floor and assume the guise of wealth. Those on the outside looking in wouldn't know if a resident lived on one or many floors.

If cost was no object, brownstone, especially from Portland, was the stone of choice for mansions, as well. In 1854 sarsaparilla king Samuel Townsend erected a four-story cube of brownstone, for some years the largest private residence in Manhattan. When George Pullman built his mansion in Chicago, he imported Portland brownstone, even though he could have used a brown sandstone from Ohio.[28] The only building on Nob Hill to survive San Francisco's 1906 quake was "Silver King" James Flood's brownstone mansion, built between 1884 and 1886 from Portland brownstone. The stone had been shipped as ballast around Cape Horn and was used to construct the 107-foot-by-127-foot residence, featuring fourteen solid stone columns, each thirteen feet tall by twenty-two inches square, and twenty-three-foot-long sandstone steps. The biggest blocks each weighed eighteen tons.[29]

Lewis Mumford termed this period the Brown Decades. He wrote in 1931 that "the Civil War shook down the blossoms and blasted the promise of spring. The colours of American civilization abruptly changed. By the time the war was over, browns had spread everywhere: mediocre drabs, dingy chocolate browns, sooty browns that merged into black. Autumn had come." Brown was a sign of "renounced ambitions, defeated hopes."[30] Like earlier critics, Mumford specifically bemoaned the East Coast's chocolate-colored sandstone.

But what damned brownstone was not its critics, it was the builders who made the mistake of not always using seasoned stone, which allowed water to infiltrate the non-case-hardened blocks. During the peak years of brownstone use builders had to take what came their way. They often received poor-quality stone, more susceptible to weathering, especially in winter, when water and ice weakened the stone during freeze-thaw cycles. After municipalities started to salt roads, the salt crystals further damaged blocks when they penetrated and started to grow in the pore spaces of the sandstone.

The second and more serious mistake builders made involved how they laid the layered stone. If a builder placed a block with its layers flat, like a book placed flat on a table, water and salt could not penetrate it. When a builder stacked blocks on edge, with bedding planes running vertically, then water and salt could force the layers to peel off the block, like what happens if you stand a book on its spine and the pages fall open. This was the phenomenon that I witnessed with the building blocks at Harvard Hall.

The use of brownstone as a curtain wall on buildings exacerbated the peeling, or spalling. On most buildings the brownstone curtain wall blocks were only four or five inches thick. Builders usually cut the blocks on site themselves and it was much easier to cut a block along the bedding plane and produce wide, thin blocks than cutting against bedding and producing tall, narrow blocks. The wide, thin blocks were the ones that peeled more often in curtain walls because the bedding ran parallel to the wall. With the tall, narrow blocks forming the curtain, the bedding was stacked perpendicular to the wall and water and salt couldn't penetrate it. When builders used brownstone for structural purposes, they cut massive blocks, on the order of twelve to twenty-four inches thick, and generally laid them correctly so the stone didn't weaken.

Poor maintenance also allowed water to seep into a block and ruin it. If an owner repointed the mortar regularly, blocks failed less often, but in the decades following the peak use of brownstone, fewer took care to do so.

No specific date marks the end of the brownstone era. Fashion was bound to change. For nearly fifty years, brownstone ruled New York. Like the Knickerbocker society that had presided over Manhattan, brownstone gave way to new ways of thinking. And just as Hitchcock's tracks were outshone by dinosaur bones, new stone arrived to make its mark on New York.

However, brownstone did not go extinct. In 1993 an ex–coal miner named Mike Meehan opened a small quarry on a ledge north of the water-filled Middlesex/Brainerd quarry. He knew nothing about quarrying brownstone. "Being a coal miner, I was more adept at blowing things up," said Meehan.[31] "But at the end of the day, I knew I wanted to be small scale and to be making a product."

He got involved with brownstone when his wife was working on a master's degree at Yale. Their kids had finished college and he sensed it was time to get out of the coal business. During one visit to Yale he saw a new building going up that used reddish sandstone. He asked the architect about the rock. He told Meehan that it cost three hundred dollars per cubic foot and that it came from Germany. "I had been selling crap for twenty-five dollars a ton," said Meehan.

Shortly thereafter, Meehan got a canoe and began to explore the Portland quarries. He looked at aerial photographs, found what looked like three quarryable ledges, and hacked his way through eight to ten feet of brush to reach them. The owner chased Meehan off the property, but not before Meehan had confirmed the quality of the rock.

He eventually worked out a deal with the owner to establish a small quarry. Meehan's first contract was for twenty-five thousand dollars' worth of stone for a restoration project at Gallaudet University in Washington, D.C. He quarried the stone and sent it to Barre, Vermont, to be cut. The only local stone the university had been able to get was from old railroad trestles. Institutions including Brown University, Yale, and Pratt Institute have continued to order his stone. In recent years, after acquiring his own milling saw, he has been providing detail pieces, such as steps and lintels, for more and more high-end homes.

He initially began quarrying the old-fashioned way, drilling long holes side by side, and then pounding wedges in to split off blocks. Later he discovered a powder, known as a silent cracking agent, that he could pour into the holes. Mixed with water the powder expands and splits the rock. More recently he acquired what is in essence a giant wire cheese slicer, except that the wire is impregnated with industrial diamonds. To cut a block, he feeds the wire down one vertical hole, fishes it out of a connecting horizontal hole, and then uses a machine to slowly pull the loop taut to cut through the sandstone. To slice slabs, he takes a block, mounts it on a cart, and wheels it under another diamond-encrusted wire. This wire travels between two spinning vertical wheels mounted about twenty feet apart on a steel frame. By lowering the wheels in tandem up and down on the frame, Meehan lowers the horizontal wire, which cuts into the block of sandstone.

"About ten to fifteen percent of each block is high-quality stone. This compares with a fifty-fifty split back in the day. We still use most of our stone but not for high-end jobs," said Meehan, who had red dust and red mud spattered on his jeans and hiking boots. To get the good stuff, Meehan cuts blocks five feet wide by ten feet long by twenty-three feet high. On one block he had marked off the twelve to fourteen inches that he will sell. The rest of the block had fractures, uneven grain size, and poor coloring.

The cut walls that ring Meehan's quarry yard look like a small scale model of the mighty sandstone cliffs of my beloved red rock country. The circular lines left by the diamond wire resemble the conchoidal fractures produced when slabs break off the Utah rock faces. Adding to the effect were the varied slabs, blocks, and boulders strewn across the yard like talus from a crumbled cliff. Meehan showed me a few blocks that have dinosaur tracks. I felt my circle had tightened further, as I finally saw in the field ancient tracks that I had previously seen only at the Amherst museum.

Meehan's brownstone blocks have also been used to restore a row house in Brooklyn. It is on a tree-lined street just north of Prospect Park, in a classic string of three-story Italianate row houses. Most have been restored, including the two on either side of Meehan's building. One is reddish brown with tan window molding and an arched doorway. The restoration on the left is army tent green. Their owners clearly put a lot of time and money into them, but compared to the building they sandwich, they look pedestrian and bland.

Meehan's new blocks have been cut to emphasize the bedding planes of the Portland rock. The beds run horizontally, in the correct direction so that they will not erode quickly. Each block is different, with variations in grain size, color, and bed thickness. They look substantial and have a warmth about them. This is the first restoration project that allows me to see the geologic history of the brownstone. It is by far the best restoration I have seen.

As I sat across the street from Meehan's blocks of brownstone, I was reminded of a comment he made at the quarry. "One hundred years from now, when people see these buildings they will say, 'That's a glorious building.' That's a good thing to me."

# 2

# THE GRANITE CITY—BOSTON GRANITE

*And nothing's as precious as a hole in the ground.*
—Midnight Oil, "Blue Sky Mine"

I WAS STANDING on what had been a big hole in the ground—a 210-foot-deep, man-made chasm two miles west of Quincy, Massachusetts. Technically, I stood on a grassy meadow, but underneath the five hundred thousand cubic yards of dirt, a few cars, tons of trash, and heaps of industrial debris, was an old granite pit.

To reach this grass-covered field, I had driven ten miles south of the tangle of roads known as Boston's Big Dig. Route 93 headed past the JFK Presidential Library and over the Neponset River to exit 8 to a nondescript off-ramp and a right turn. At a small, unmarked parking lot about a mile up, I followed a paved path by a pile of empty Budweiser cans and onto the green mound.

Granite walls ringed the field. One on the right sloped steeply up to a flat ledge and one directly opposite rose seventy-five feet high to a prowlike point, known appropriately as Ship's End. Between it and another towering wall of granite to the left, cattails grew out of an ice-covered pool. The dominant human influence was graffiti, which so densely covered the walls in shades of red, blue, yellow, and black that it obscured the underlying rock. Someone named Joyce, or one of her fans, appeared to have had a good supply of green paint; at least five shamrocks emblazoned with her name dotted the granite.

Atop the thirty-foot-high slope to the right of the field, iron bolts, deadman anchors, eye rings, and rods stuck out of the rock like some slightly out-of-whack Arthurian challenge. Additional iron bits jutted from a waterbed-sized granite block—the base of a derrick—cemented to the granite surface. There was evidence of derricks at other high spots, as well as beer cans, broken bottles, fire rings, trash, and clothes. In the distance, from one high point, rose the tall buildings of Boston.

Wandering along the granite walls and then through the quiet meadow, it is hard to imagine that this field covers one of the great industrial sites on the eastern seaboard. Out of this now-filled-in hole, known as the Granite Railway Quarry, came the gray granite used to build the U.S. Navy's massive dry docks in Charlestown, Massachusetts, and Norfolk, Virginia; the great customhouses in New Orleans, Boston, and Mobile, Alabama; and most important, the Bunker Hill Monument. This 450-million-year-old granite led, in 1826, to the first commercial railway in the country, and it was the Granite Railway Quarry that made granite *the* quintessential building stone of the early and middle 1800s. The popularity of this quarry eventually led to fifty-three additional quarries opening around Quincy and gave the town its moniker, the Granite City.

These quarries west of Quincy (pronounced *Quin*-zee) operated from 1826 to 1963. By 1838, one-sixth of the town's residents worked in the vast holes. They came from Finland, Italy, Ireland, Scotland, Germany, Sweden, and Canada. It was dangerous and debilitating work. The constant drilling left many of the workers partly deaf. Steel cables and iron chains snapped and whipped into the men as they dug. Blocks of rock fell from derricks or exploded from the walls. Derricks collapsed and workers got crushed. Men froze in winter and baked in summer. They breathed toxic dust.

Appalling conditions aside, the quarries must have been an awesome sight during their heyday. Men clambered over walls of debris to ledges where they cut deep grooves with hammer and chisel or pneumatic drills. They pried out enormous cubes of granite and attached them to long cables dangling from derricks that dotted the quarry's rim. Other groups of men transported the titanic masses down to a railroad that took the blocks to finishing sheds. And the entire time, pumps sucked out the water that wept continuously from the quarry walls and threatened to flood the twenty-story-deep orifice. Now, all that remained of the Granite Railway Quarry was this bucolic setting of grass, graffiti, and granite.

★ ★ ★

In his magisterial *A History of Old Braintree and Quincy*, William Pattee wrote that "the first building of any architectural pretensions" made from Quincy granite was King's Chapel, a small church completed in 1754.[1] Now dwarfed by more recent granite, steel, and brick structures, King's Chapel is the oldest extant granite building in Boston, but not the first building to incorporate granite. As early as 1650, builders had used granite in structures but mostly for foundations, lintels, and sills. Not until John Hancock's uncle Thomas built a brownstone-accented, masonry house in 1737 was one made with solid granite walls.

Nor is King's Chapel the first use of stone on this site in downtown Boston. Slate gravestones, many of which are slowly sinking back into the earth, fill the cemetery next to the chapel. Established in 1630 and formerly the town burying ground, the cemetery is the resting place for both the famous—John Winthrop, first governor of the Massachusetts Bay Company—and the forgotten—William Dawes Jr., one of the other two people who alerted colonists that the "British are coming" on that fateful night of April 18, 1775. But King's Chapel is clearly the principal stone edifice of the block.

Architect Peter Harrison designed what many consider to be the finest example of Georgian church architecture in America. Twelve painted wood Ionic columns, six in front and three on each side, support a plain entablature topped by a railing. A flat-roofed tower with an arched window rises from ground level and extends one story through and above the railed portico. The main body of the chapel is simply a rectangle with a hipped roof. With its linear, symmetrical design, King's Chapel is the type of building that probably gives architect Frank Gehry nightmares.

Take a closer look, however, and the chapel reveals an enchanting complexity, at least in its building stones. Not only are the blocks a mix of colors, ranging from oatmeal through ash to dark gray, but also a mix of sizes, from stubby to long and lean. This diversity gives King's Chapel a more organic feel than modern stone-veneer buildings where uniformity reigns. In addition, hammer marks have left the stones' surfaces stippled, as if the little church has goose bumps.

Heterogeneity was not necessarily a design feature, but a result of how the blocks were quarried. The King's Chapel granite came from boulders (or bowlders, as it was often spelled in the 1700s) scattered on the ground in Braintree, a mile or so from what would become the Granite Railway Quarry. Masons split the stone by building a fire on top of a boulder, which weakened the rock. The men then dropped iron balls, called bee-

King's Chapel, Boston, built 1754.

tles, to crack the boulder into rough blocks. Additional hammering and chiseling squared the stones.

In an 1859 speech describing the history of building stone in Boston, Massachusetts, Chief Justice Lemuel Shaw said "When this work [King's Chapel] was finished, it was the wonder of the country round. People coming from a distance made it an object to see and admire this structure . . . But it seemed to be universally conceded, that enough [stone] more like it could not be found to build such another."[2]

Yet only a hundred years after the erection of King's Chapel, when Justice Shaw made his speech, granite had become commonplace in Boston. The change occurred because of a carriage ride in 1803. Massachusetts lieutenant governor Edward H. Robbins, a member of the commission in charge of building the Charles Bulfinch-designed State Prison in Charlestown, was traveling through Salem when he noticed a building made of rock cut in a manner new to him. On the basement walls, he found tool marks in the rock spaced about seven inches apart. Since one of Robbins's goals was to find better cut-stone prices for the prison, he stopped and asked for the owner of the building. He referred Robbins to the contractor, who told the lieutenant governor that a Mr. Tarbox had provided the stone and that he lived in Danvers, two to three miles away.[3]

Robbins headed to Danvers and found Tarbox living in poverty in a small house. With prodding from Robbins, Tarbox described his method of cutting stone. First, he drilled a line of holes into the rock. He formed each hole by hammering a sharpened steel rod, or drill, turning the drill, and cleaning the hole with a specialized spoon. He hammered, turned, and spooned until he reached a depth of two to four inches. He then inserted into each hole two metal, half-round shims, each bent at the top to prevent them from slipping into the hole. Between the shims, known as feathers, he placed a square metal wedge, the plug. To separate the rock, he pounded the plugs in succession until a slight cracking noise alerted Tarbox that the stone would fail on its own.

Flabbergasted by what became known as the "plug and feather method," Robbins asked Tarbox if he would travel to Quincy and teach the local stonecutters this new and novel technique. Tarbox replied that he couldn't go because he had a family to feed and no proper clothes to wear, whereupon Robbins offered to take care of Tarbox's family, pay him two to three times the wages of the best stonecutters, and go into Salem and buy him proper clothes. Within three months every stonecutter in Quincy and nearby Braintree cut stone in the Tarbox method.

Others, however, claim that the credit for inventing the plug and feather method should go to three Quincy men, Josiah Bemis, George Stearns, and Michael Wild. In his history of Quincy, William Pattee wrote that in 1803 the trio split stone with plugs and feathers. "So elated were these gentlemen on this memorable Sunday, that they adjourned to Newcomb's Hotel, where they partook of a sumptuous repast."[4]

The plug and feather method was a radical improvement over the fire and hammer application previously used by Massachusetts stonemasons, but neither Tarbox nor the well-fed Quincy men invented the technique. In 1757 one of my heroes, America's first great naturalist, John Bartram, described in a letter how he built four houses with a plug and feather method. Bartram wrote that with this technique "you may split them as true almost as sawn timber."[5]

Egyptian builders, however, predated Bartram by at least four thousand years. They either inserted wood wedges in the holes and pounded the wood or soaked the wood with water and let the expanding wood fracture the stone. Finnish stonemasons employed a parallel method but one more suited to a northern climate. Late in the year, they filled drill holes with water and capped them tightly. When the water froze it expanded and cracked the stone along the line of holes. An 1889 report noted that this process separated four-hundred-ton blocks.[6]

Justice Shaw, however, credits Tarbox alone as the creative genius responsible for his city's recently hewn granite face. Despite his august accomplishment, no one seems to know Tarbox's first name. Neither do we know when he was born nor when he died. We do know, however, that within a few months of Tarbox's journey to Quincy, the price of cut stone dropped 40 percent, setting the stage for the Granite Railway Quarry and ultimately a fundamental change in the look of cities across the eastern seaboard.

One of my most pleasant discoveries when Marjorie and I moved to Boston in 1996 was the prevalence of granite as a building stone. I had come east expecting the worst. I was leaving twelve-thousand-foot-high mountain ranges and towering cliffs of sandstone for a landscape whose only geologic feature that I knew of was a boulder that a boatload of religious renegades supposedly stepped onto in 1620. When we got here, though, I found granite curbs, benches, and bridges, as well as a beautiful pink granite library and gray granite post office and museum.

Granite also crept up in friends' houses. Granite countertops have become one of the hottest features in real estate. If you look at ads for any condo or town house or new subdivision, most listings trumpet the stone. The popularity of these must-have kitchen accoutrements has helped make granite perhaps the best-known type of stone to nongeologists. (Not forgetting, of course, how every four years granite bursts on the scene in the form of Olympic curling stones.)

Throughout the country, granite is widespread and common. It tops the lower forty-eight's highest spot, Mt. Whitney, and floors the deepest gorge, Hells Canyon on the Snake River. Granite is a type of igneous rock that began as a liquid, or magma, and solidified within the planet. Granites appeared at the surface only when tectonic forces shoved them up or erosion stripped away the overlying rocks. The word "granite," comes from the Italian *granito*, or grained, in reference to the interlocked grains of minerals that make up the stone.

In a hand specimen and particularly in the wall of a building where they have been polished, granites resemble no other building stone. Minerals range in size from a peppercorn to a plum. Some minerals have a glassy appearance. Others twinkle. Most are dull. Granite can be pink or red, infrequently green to black, and commonly white to gray. Rarely will you find any layering, consistent orientation of the grains, or swirls. You will never find fossils.

The Quincy Granite fits this general description, albeit with idiosyncrasies. Most granites contain two types of the mineral feldspar, broadly called plagioclase feldspar and alkali feldspar. In contrast, Quincy contains only alkali feldspar, a result of solidifying at a high temperature. Alkali feldspar gives the rock its characteristic green-tinted, dusky gray color. (One Quincy quarry owner called himself the "Extra Dark Man" because of the particularly dark stone excavated from his property.) Further darkening results from Quincy's nearly black quartz, as opposed to the more common clear or white varieties.[7] High temperature also sapped the Quincy Granite's magma of another typical granite mineral, mica, and led to the formation of an unusual mineral known as riebeckite. Because riebeckite is harder than mica, it allows the Quincy Granite to take a high polish. Riebeckite also contains a very stable form of iron, which means that Quincy Granite doesn't rust and stain when it weathers.

Dark and hard, polishable, and weather-resistant, and with Tarbox's plug and feather cutting technology in place, the Quincy Granite was an ideal stone for the incipient building trade. All it needed was a signature building.

Little was done with granite for the twenty years after Tarbox ventured south to Quincy. Workers completed Bulfinch's State Prison in 1805, although not with Quincy Granite. Instead, Bulfinch used Chelmsford granite, which could be floated twenty-six miles down the recently completed Middlesex Canal. Granite also went into the Boston courthouse, University Hall at Harvard, several Boston churches, and Massachusetts General Hospital, but these, and a few others, were the only notable granite buildings to appear by 1825.[8] Although made from granite, all were built in a traditional style using conventionally sized blocks.

This interlude between 1803 and 1825 was granite's incubation period. Architects were experimenting with how to design with granite, and transportation was a problem. In Quincy, stonemasons were learning better how to take advantage of the plug and feather technique and still primarily working with rocks they found on the ground. What would become the Granite Railway Quarry was still a forested knob known as Pine Hill when incubation ended.

In 1825 an architect-engineer named Solomon Willard arrived in Quincy. Legend has it that he had walked three hundred miles across New Hampshire, Maine, and Massachusetts in search of the perfect granite for what would become his most famous building, the Bunker Hill Monument. Willard found that granite at a ledge in a wooded area about a mile from the King's Chapel "quarry" site.[9]

Born June 2, 1783, in Petersham, Massachusetts, Willard spent his youth on the family farm and in his father's carpentry shop. He moved to Boston when he turned twenty-one "to seek, not his fortune (as is the object of so many), but his own intellectual improvement, and the means and opportunity of doing greater good," according to the official history of the Bunker Hill Monument.[10]

When not helping his fellow man, Willard studied architecture and drawing. He relied on carpentry to make a living. Willard carved columns for the steeple at Park Street Church, an eagle for the pediment at the old Custom House, and a model of the U.S. Capitol for Charles Bulfinch. He began carving stone and in 1819 cut the Ionic columns for St. Paul's Church, which launched the fad for Greek Revival structures in Boston. These columns were made from Aquia Creek sandstone, a notoriously crumbly rock quarried in Virginia and used in the U.S. Capitol and the White House. Willard also began to design buildings and to teach drawing and architecture. By the early 1820s, he was well known and well regarded in Boston.

His involvement with Bunker Hill began in August 1824, when directors of the Bunker Hill Monument Association (BHMA) asked Willard to submit a plan for a monumental column. Boston Brahmins, such as Daniel Webster and Thomas Handasyd Perkins, had formed the association in 1823 to build "a simple, majestic, lofty, and permanent monument, which shall carry down to remote ages a testimony . . . to the heroic virtue and courage of those men who began and achieved the independence of their country."[11]

By 1825 the BHMA had raised money and decided what type of monument would best honor one of the most important battles of the Revolution, which the American forces lost and that didn't occur on

Bunker Hill. They also purchased property—eventually totaling fifteen acres—on Breed's Hill, where the battle did take place. Apparently not pleased with Willard's potential design, the directors also published a notice in Boston papers and around the country announcing a design competition, with the winner receiving a hundred dollars. In response, Willard told Boston's best and brightest he "had no wish to enter into any contest about the designs."[12] (I suspect that in private his words were a bit saltier.)

Fifty designs were submitted, including an obelisk from Robert Mills, who designed the Washington Monument in 1836. A former student of Willard's, Horatio Greenough, won the competition in April with his own obelisk. In his memoir Greenough wrote, "The obelisk . . . says but one word, but it speaks loud. If I understand its voice, it says, Here! It says no more."[13]

Despite Greenough's plan, the directors dithered on whether an obelisk or a column was more appropriate for the monument. Part of the problem was style and part was cost, so the BHMA appealed to Willard to make a cost estimate for an obelisk and for a column. After learning that a column would cost $75,000 and an obelisk $60,000, the directors finally reduced their decision to one all could agree upon and chose the obelisk. Keeping to their rapid-fire decision-making, they appointed a committee of five to prepare a design.

It was most likely during his cost preparation work that Willard found his ledge in Quincy. Two days after the directors' meeting, a young engineer and friend of Willard's, Gridley Bryant, who had helped Willard on the cost estimates, bought the four-acre property in Quincy for $250.[14]

Ten days after approving the obelisk, and for no apparent reason not willing to use Greenough's plan, association directors laid a cornerstone commemorating the fiftieth anniversary of the Battle of Bunker Hill. Thousands watched on June 17, 1825, as the great French general the Marquis de Lafayette led the ceremonies. Workers buried a box within the cornerstone that contained official accounts of the battle, coins and medals, and a piece of Plymouth Rock, a 620-million-year-old granite from Dedham.[15]

The design committee presented their plan on July 5. Taller and simpler than Greenough's obelisk, the new proposal included details on the foundation, the interior lighting and steps, size of stones, obelisk dimension, and cost. Construction and design now totaled $100,000. The committee also recommended hiring an architect and superintendent.

Continuing to move at their normal speedy pace, the directors named Solomon Willard as architect and superintendent on October 31, 1825. He offered to do the job for free but ultimately accepted the $500 per year the committee insisted he be paid. He also gave $1,000 to the BHMA. His first proposal as superintendent was to purchase the ledge of granite from Gridley Bryant, which the association did for $325, a nice profit for Bryant. During the winter Willard finalized the drawings for the obelisk, increasing the size of the building blocks from eighteen inches to thirty-two inches tall.

Transportation presented the central snag for Willard. How would he move blocks that weighed up to six tons across the twelve miles of swamp, forest, and farms that separated Quincy from Charlestown? Willard favored either a completely overland route or moving the stone in winter, when sledges could carry the blocks to the Neponset River, four miles north. A barge would transport the stone through Boston Harbor to Charlestown, which formed a peninsula on the north side of the Charles River, due north of downtown Boston. Gridley Bryant had another idea.

Six years younger than Willard, Bryant was born in Scituate, Massachusetts. In recalling his childhood, Bryant wrote that he had a "mechanical and inventive turn of mind . . . I was generally at the head of the young urchins of our neighborhood, and when there was a fort to be constructed . . . I was always appointed the chief engineer." Despite his friends' high regard, his mother pushed him out of the house at fifteen to apprentice with a leading Boston builder. Six years later he headed out on his own and by the early 1820s, Bryant was one of the foremost masons in Boston, including a stint working for Willard.[16]

In late 1825 Bryant suggested a railway as the best means to transport granite from the quarry to the Neponset River. He came up with the idea after hearing about English railroads transporting stone from quarries. The BHMA rejected his plan as too ambitious. Not to be thwarted, Bryant presented his idea to several businessmen, including Thomas Handasyd Perkins, who although a member of the association, was open to Bryant's proposition.[17]

Perkins, a Boston merchant and philanthropist who had made a fortune in the China trade, not the least of which was from opium, endorsed the project. He knew about railroads from trips to England and recognized their moneymaking potential. With Perkins's prompting, a group made a petition to the state legislature on January 5, 1826, to establish a railroad. The bill passed on March 4 chartering the Granite Railway Company with Perkins as president. (In a letter, Perkins wrote "I think I may safely call it my road, not only because I set it agoing, but because I own 3/5ths of it.")[18] The Perkins group hired Bryant as superintendent and designer. On April 1, 1826, he broke ground on what was called, and is often still called, the first railroad in the United States.

"If you use enough adjectives, you can get it right," said Vic Campbell, who has spent the past forty years researching the history of the Granite Railway, locating its route, and telling people about it.[19] "The Granite Railway was the first chartered, commercial railway in the United States." He noted that Fred Gamst, a former anthropology professor at the University of Massachusetts, determined that the Granite Railway was actually the twelfth American railway. The first was in Boston, ran about one-quarter mile, and carried dirt off Beacon Hill to fill the Back Bay. Built in 1805, the Beacon Hill Railroad would have been running when a displaced Scituate teenager arrived in town. Gamst speculated that such an unusual operation would have attracted the attention of a young Gridley Bryant despite his later claim that "all the cars, trucks, and machinery are my original invention."[20]

Campbell stood just below the site of Willard's ledge, or what was left of it, since any rock that enticed Willard is now twelve miles away in the monument. A ledgy granite hill covered, probably as it was in 1825, by pine trees and boulders, rose above the mostly flat and bare ground. As Campbell walked among piles of cut and uncut granite, he pointed out where the masons squared and numbered stones, the blacksmith's shop, and the superintendent's building.

He stopped near several large rectangular stones partially buried in the soil. "This is mile zero of the Granite Railway," said Campbell. About 125 feet of moss-covered path ran along the ground. Every eight feet a ten-inch-square, seven-foot-long granite block, known as a sleeper, lay perpendicular to the path. This is the only relatively intact section left of the Granite Railway, although no rails remain.

The sleepers rest on several feet of crushed rock, which Bryant used to protect the tracks from frost-induced buckling. He then placed pine rails, six inches wide by twelve inches high, on top of the sleepers. The tracks were five feet apart and held in place by iron pins, several of which still stick out of the granite. On top of the pine ran four-by-four-inch oak timbers supporting a quarter-inch-thick iron strip, called a strap rail, where the flanged wheels rested. When the wood wore out, Bryant replaced it with granite. As we reached the end of the path, which extended into a shrub-filled swamp between rock walls, Campbell's enthusiastic gestures nearly propelled him off the raised railroad, but I grabbed him before he tumbled off the six-foot-high wall.

Bryant's most innovative design was his rail car, fourteen feet long, eleven feet tall, and supported by six-and-one-half-foot-high wheels. The empty car would back up to where the cut blocks were. Workers would turn gears on the car, which would lower a pallet supported by six chains. They would unhook the pallet, move the car forward, load a block or blocks, and back the car up again. "One man could raise a six-ton block, which could be up to three feet wide and thirty-two inches high," said Campbell.

After crossing on trestles over the swamp, the railroad curved around Pine Hill and by Willard's house, before heading straight northwest to the Neponset River. Along the way it dropped eighty-five feet. Two pullouts allowed cars going in opposite directions to pass. Not that traffic was much of a problem but the route did cross several streets, where chains prevented slow speed collisions. The railroad ended at a twelve-hundred-foot-long wharf, which took six months to build and cost two-thirds of the total fifty-thousand-dollar price of the railroad.

Barges and sloops carried the blocks down the Neponset, across Boston Harbor to Charlestown. They landed at Deven's Wharf, coincidentally near where Paul Revere began the most famous horse ride in American history. Subsequent filling in of the harbor has obliterated the wharf, which was adjacent to the Charles River Bridge. Ox-drawn carts carried the blocks the final few hundred yards up to the building site.

Bryant made the first test run of the railroad on October 7, 1826. Workers loaded three cars with sixteen tons of rock and a single horse pulled the entire load. The horse arrived at the Neponset wharf in less than an hour. The railroad men celebrated, drinking from brown glass bottles imprinted with "Success to the Railroad."[21] The next load would not arrive at the river until the following spring.

"Willard was never a big supporter of the railroad. He wanted the work done faster," said Campbell. Willard also worried about stone breakage on the railroad and paid for a survey of the twelve-mile overland route. "He started work on quarrying in March 1826, built up a huge collection of blocks, and had to wait until March 1827 before the railroad carried any stone. He stormed off the job several times in the first few years because of his concerns," said Campbell.

"The Quincy Granite is its own special case," said Dick Bailey, a geologist at Northeastern University.[22] He has spent more than three decades studying New England geology, focusing primarily on the Avalon terrane, a suite of rocks central to understanding the oddball history of the Quincy Granite.

A terrane is a fault-bounded body of rock, with limited extent, characterized by a geologic history different from the history of nearby rocks. Current thinking holds that the Avalon terrane began life as part of South America. The central line of evidence comes from chemical analysis of minute minerals, called zircons, found in both northwest South America and in rocks around Boston. Bailey rejects this model. He favors an African origin.

His main evidence is a truly astounding creature, a trilobite fossil discovered in 1834 at a slate quarry three miles east of the Granite Railway Quarry. The most charismatic crustacean of the Cambrian Period, trilobites foraged on ocean bottoms for 300 million years. Most species were a couple of inches long, or shorter, and resembled a pill bug, or rolypoly. The one from Quincy, *Paradoxides harlani*, was twelve inches long, which led Percy Raymond, a curator at Harvard's Museum of Comparative Zoology, to write in 1914 that *Paradoxides* is "such an oasis in the sterility of Massachusetts paleontology" that it borders on the "domain of romance."[23]

Raymond didn't know the most exotic part of the story, that the 510-million-year-old *Paradoxides* was not a North American trilobite. Its closest relatives, including a sixteen-inch-long Moroccan giant, lived in north Africa. Along with several other much smaller trilobites from the same quarry, the Quincy fossils provided the first clue that Avalon was an exotic landmass. "I am a paleontologist, which is why I haven't given up on the northwest Africa position. There's something a lot more compelling to the trilobites than to isotope analyses," said Bailey.

Avalon's plate tectonics history can best be described in accordionlike terms with continents opening, oceans closing, oceans opening, and finally landmasses colliding. The terrane's story began between 700 and 800 million years ago on the edge of a supercontinent known as Rodinia

(Russian for motherland). Consisting of parts of Gondwana and Laurentia (North America and Greenland) that had glommed together 300 million years earlier, Rodinia was beginning to rip itself apart. A massive rift valley formed and filled with nearly pure quartz sands carried by rivers. Known as the Westboro Formation, the sands later lithified into what is New England's oldest sedimentary rock.

The next oldest rocks formed 620 million years ago when a closing ocean basin slammed a narrow belt of islands into the continent. This collision produced a pink granite that resembles Ben and Jerry's Cherry Garcia Ice cream. Known as the Dedham Granite, it is geologically the oldest Boston-based rock used as building stone. Dedham Granite is also what the Pilgrims first stepped onto when they landed in 1620.

As another rift opened, sediments began to flow into a marine basin. This collection of rocks has been designated by geologists as the Boston Bay Group. Bailey's work has helped show that more than seventeen thousand feet of mudstones, sandstones, and conglomerates accumulated in the sea that dominated the area between 605 and 543 million years ago.

The Boston Bay Group includes two other important local building stones. The first is Cambridge slate, one of the earlier stones in the building trade. The second building stone is a purple-hued stew of boulders, cobbles, and pebbles suspended in a fine-grained mud. Used in over thirty-five Boston churches and known as Roxbury puddingstone, it was celebrated by Oliver Wendell Holmes in his 1830 poem, "The Dorchester Giant."[24] He attributed the odd stone composition to the giant's three unruly children and a "pudding stuffed with plums."

*They flung it over to Roxbury hills,*
*They flung it over the plain,*
*And all over Milton and Dorchester too*
*Great lumps of pudding the giants threw;*
*They tumbled as thick as rain.*[25]

After the children finished their roughhousing, Percy Raymond's great trilobite appeared, roughly a half billion years ago. The foot-long crustaceans crawled around the shallow water of the Iapetus Ocean (Atlas's father in Greek mythology), off the edge of the small Avalonian landmass. This microcontinent, akin to modern-day Madagascar, lay just west of Gondwana, far south in the southern hemisphere. And then Avalon, and its cache of trilobites, abandoned its point of origin near Africa and began to travel north on the plate tectonics highway.

Around 450 million years ago, Avalon passed over a hot spot, or zone of weakness in the earth's crust where heat escapes from the mantle, as now occurs in Hawaii and Yellowstone. This fiery Bunsen burner melted surrounding rock and generated a hot, dry alkali-feldspar-rich magma that punched its way into a trilobite-rich sedimentary rock. The Quincy Granite was born, although no one knows exactly where or when, noted Bailey.

Avalon continued to glide north and west through the Iapetus, probably jostling with other terranes before docking on the eastern edge of North America sometime between 425 million and 370 million years ago. Another terrane rammed Avalon between 300 and 250 million years ago, followed closely by Africa and Europe. This was the final squeeze of the accordion and marked the penultimate stages of the closing of the trilobite-rich Iapetus Ocean and formation of Pangaea. The assembly of North America, at least in the east, was more or less over. Geologic quiet has dominated the east ever since, with a few big events, most notably the breakup of Pangaea, which started the continents on their journey to their modern locations.

Because of this geologic calm, the Quincy Granite did not get beaten up. By not experiencing the trauma of continental collisions, great uplift, and weathering, or deep burial under later sediments, the Quincy Granite lacks joints and other zones of weakness and can form massive, magnificent blocks, which in 1825 had attracted the attention of Solomon Willard, the gallivanting Boston architect-engineer.

Work began again on Bunker Hill Monument in April 1827, nearly two years after General Lafayette had placed the original cornerstone.[26] (The box containing the bit of Plymouth Rock was reburied in a new cornerstone in the northeast corner.) One of the first things Willard had to do was to figure out how to lift the massive blocks. Working with local seaman Almoran Holmes, Willard designed a massive hoist, or derrick, which consisted of a single wooden pole and a movable boom, attached by a block and tackle. Cables facilitated rotation of the boom, changing its angle, which determined reach. Known as the Holmes Hoisting Apparatus, it had a reach of fifty feet and, with six horses providing power, could lift up to twenty tons. The derricks along the edge of the Granite

Railway Quarry were based on Holmes's design. Ironically, Holmes died several years later while using one of his own derricks.

With Bryant's railroad transporting blocks and the derrick lifting stones, workers were able to lay fourteen courses of the monument, a height of thirty-seven feet, by February 1829.[27] And then the Bunker Hill Monument Association's money ran out, work stopped, and all employees, including Willard, were laid off. During the work break, the Granite Railway's owners decided to open a new quarry on land about

Workers at Granite Railway Quarry, 1923.

a half mile closer to the Neponset than Willard's quarry. That land, which Perkins and his fellow railroad owners had purchased in 1826 for ten thousand dollars, became the Granite Railway Quarry.

When work on the monument started again in 1834, the Granite Railway Quarry supplied the rock. Money ran out once more in November 1835, after the addition of eighteen courses. The obelisk-to-be stood at eighty-five feet.[28]

Construction did not begin again until May 1841. In the intervening years, the association sold ten of their fifteen acres to raise cash.[29] The final money came from a fair held in Boston in September 1840. Organized by the "inspiring influence and delicate hands of the gentler sex," the fair netted over thirty thousand dollars on sales of a "variety of things to please the eye, to adorn the house or person, or to supply the common wants of life."[30]

Workers placed the final stone on the monument at six A.M. on July 23, 1842. A formal dedication took place on June 17, 1843, with 110 Revolutionary War veterans present, including ninety-seven-year-old Phineas Johnson, who had fought at Bunker Hill sixty-eight years earlier. The cost to build the monument was $101,680, basically on budget.[31]

The best way to see the great obelisk is to follow the Freedom Trail from downtown Boston across the Charles River to Charlestown. Designated by a red line, either painted or made of brick, the trail winds for three miles through Boston and highlights many Revolutionary War sites, such as Paul Revere's house and the Old North Church. After crossing the river, the red line heads up a small hill, bordered by a wind tunnel of brick, wood, and granite row houses.

The narrow street leads to an open square at the top of the hill and more row houses. A final short flight of stairs enters the monument grounds, where you can look back at Boston and see the recently built Charles River Bridge, whose 270-foot-tall, cable-support towers were designed to look like the monument. Greeting you is a statue of Colonel William Prescott, who famously warned his men, "Don't one of you fire until you see the whites of their eyes," just hours after he had set up his defenses on the wrong hill; Bunker is higher and better situated for controlling Charlestown than Breed's Hill. With a sword in his right hand and his left hand trailing behind warning his men to wait, Prescott stands ready to take on any soldier or park visitor.

Behind Prescott and a metal gate towers the 221-foot-5-inch-tall monument. Its Quincy Granite building blocks are immense. Over 13.2

million pounds of stone make up the obelisk. The biggest stones measure 32 inches high by 90 inches long and weigh up to five tons. Unlike the King's Chapel blocks, the Quincy Granite stones are smooth and matching in color, gray with a few dark streaks. Outside of lightning rods that run up two corners of the obelisk, no other ornamentation mars the simple structure.

To access the monument, you go into the visitor center, exit by a side door, and enter the obelisk. Spiraling up around the central column are 294 steps. In winter, water dripping from granite can create long icicles, whereas in summer, the monument is pleasantly cool. A handful of narrow windows—which occasionally contain birds' nests—bring in both light and air. An open room made by the pyramid at the top has four small square windows that provide an unparalleled view of Boston and the surrounding area.

Even before completion of the monument, its construction, as well as the development of the Granite Railway, led to granite finally becoming the preeminent building stone in Boston. Willard showed that large blocks could be used and transported, and by refining quarry techniques, he helped drive the price down by 75 percent.

Willard's work became "the standard for public building in Boston—monumental, severe, and permanent," wrote art historian Jane Holtz Kay in *Lost Boston*.[32] Designated the Boston granite style, the buildings were often massive, such as the 535-foot-long Quincy Market or Boston Custom House with its forty-two-ton columns that took fifty-five oxen and twelve horses to pull. Many of these early structures still bear the perforation marks of Tarbox's plug and feather technique.

As Boston grew in prominence, its leading architectural style spread. Customhouses in Savannah, Georgia; San Francisco; and Portland, Maine, used Quincy Granite. In 1836 Willard provided stone for the New York Merchants Exchange, designed by another former student, Isaiah Rogers. Quincy quarrymen also shipped millions of paving stones to New York (still visible in a few streets in Greenwich Village, SoHo, and TriBeCa), Philadelphia, and Washington, D.C. (Vic Campbell said that one of the major paving stone suppliers didn't actually quarry the stones, they simply collected waste from a quarry on the hill above their shop and cut it into paving blocks.) The hole in the Granite Railway Quarry began to grow.

Other granite areas beside Quincy also prospered. Rockport quarries provided a dark gray rock, transported by sea. The pink Milford Granite ended up in the Boston Public Library and New York's Penn Station, and granite from Chelmsford was floated down the Middlesex Canal to the state prison, where it was cut. Rocks were shipped around the country and to Cuba.

Each of the New England states began to excavate, cut, and ship granite as well. By 1889, Maine had 153 granite quarries, including one in Vinalhaven that employed fifteen hundred people. Quarries in New Hampshire (now nicknamed the Granite State), Rhode Island, Connecticut, and Vermont (home of Barre, the self-proclaimed "Granite Center of the World") generated everything from paving stones to a single block three hundred feet long, twenty feet wide, and six to ten feet thick.

They all shared one characteristic—access to transportation. In Maine the quarries were situated on the coast. Vermont's were near railroads, as were Connecticut's, which also had quarries on rivers and the coast. In areas where transportation was a problem, builders used the stone locally and moved rock via carts over dirt or cobblestone roads.

Granite continued to dominate as a building stone for many years, aided by its physical attributes. Because of the abundant feldspar and quartz, granite is twice as hard as limestone or marble, up to twice as hard as slate, and at least equal to and generally harder than sandstone, which is usually also made of feldspar and quartz grains. Because of the interlocked minerals, granite is significantly less porous than sandstone and limestone, and about equal in porosity to marble and slate. In the age before steel beams, when stone had to provide the only means of support, granite's compressive strength made it essential for monumental structures.

In the past two decades granite has again become popular, for many of the same reasons, although compressibility is less important with steel infrastructures. Homeowners desire it because of its hardness. (Slicing and dicing with a good sharp Henckels knife will scratch a marble countertop.) Now, however, granite from the United States has lost ground to that of Finland, China, Norway, Sardinia, and South Africa. Baltic Brown, Big White Flower, Blue Pearl, Rosa Beta, and Zimbabwe Black are some granites that now dominate the market.

The Quincy Granite's massive nature, as well as its dark color and ability to take a polish, helped make it a popular building stone until about the Civil War. As railroads spread, however, less expensive granite started to flood the market and undercut Quincy's competitiveness. Devastating fires in Chicago and Boston further weakened demand by revealing that heat flaked and cracked granite. A second wave of demand for Quincy did rise in the 1880s and 1890s but not as a building stone. Instead, people wanted the dark granite for Civil War monuments and later for gravestones.

The beginning of the end of the Quincy quarries began in World War I when people stripped them of iron and steel and scrapped the machinery to melt down for shipbuilding. Demand continued to drop and finally plummeted during the Great Depression. World War II sealed the industry's fate, and in 1942 the Granite Railway Company folded. The final large quarry, Swingle's, limped along until 1963.

With the industry shut down, the Granite Railway Quarry and other nearby quarries filled with water. They became notorious as unwatched places for dumping cars, trash, and the occasional dead mobster. At least thirteen people died swimming, diving, and climbing in them. In the late 1990s, the police officially declared the Granite Railway Quarry a crime scene while divers searched for a young woman who they suspected had been murdered and dumped in the deep water. Instead, they located an Irish teenager, who had been missing for three years. Divers thought they saw another body but never found it despite draining the quarry. With the quarries finally dry, the land owners, the Metropolitan District Commission (MDC), decided to fill the hole and prevent water from seeping back in.

Coincidentally, workers in Boston were digging an even larger hole and needed to dispose of their dirt. The MDC gladly accepted five hundred thousand cubic yards of clay and lightly contaminated soil from the Big Dig, as well as seven hundred thousand dollars in tipping fees. An additional 12 million tons of Big Dig dirt were used to fill in other quarries around Quincy and to make a twenty-seven-hole golf course, four Little League fields, two soccer fields, and luxury homes, complete with granite countertops. By 2002 the Granite Railway Quarry was safe, grass was growing, and 174 years of history were buried.

# 3

# POETRY IN STONE—CARMEL GRANITE

*Here on the rock it is great and beautiful, here on the foam-wet granite sea fang it is easy to praise*
*Life and water and the shining stones.*
—Robinson Jeffers, "Meditation on Saviors"

A CENTURY LATER, at the opposite end of the continent from Bunker Hill Monument, another transformation occurred because of granite. This time, however, the stone affected just one man—the poet Robinson Jeffers. The granite so infused his life that it helped transform him from an insecure, mediocre writer to one of the great American poets of the twentieth century.

Jeffers used granite to build his private residence and a forty-foot-high tower. He called the structures Tor House and Hawk Tower and referred to the granite as "sea-orphaned stone."[1] The rock came from the beach below his house, which stands on a low hill that rises from the Pacific Ocean in Carmel, California. Jeffers placed each granite boulder by hand, generally in the afternoon after he had spent the morning working on his poetry.

In describing the changes in him, Jeffers's wife, Una, wrote, "As he helped the masons shift and place the wind- and wave-worn granite I think he realized some kinship with it and became aware of strengths in himself unknown before. Thus at the age of thirty-one there came to him a kind of awakening such as adolescents and religious converts are said to experience."[2]

I first saw Tor House and Hawk Tower in 2002 from the road that runs along the water below them. Light green grasses, gray green shrubs and a few light gray boulders covered the slope leading from the road up to the stone buildings, behind which stood a row of wind-shaped Monterey cypresses. The house was squat with a narrow line of windows just below a small triangle of brown roof. The tower was square, about half the width of the house, and topped by a square turret with two eyelike windows opening out to the ocean. The structures didn't appear to have been built so much as to have emerged geologically from the hillside, as if Jeffers had used the nearby cliffs, seastacks, and outcroppings for blueprints.

Hawk Tower, built in 1920–1925 by Robinson Jeffers, Carmel, California.

Up close, the buildings sustained my first impressions of geology manifest as home. No two stones were alike and rarely did stones of the same size rest next to each other. Edges were not perfectly straight but looked weathered and eroded. Barnacles still covered some of the stones Jeffers liberated from the sea. Finger trails ran through the mortar, trace fossils of a man and his passion.

Jeffers mixed small, large, and immense boulders in a planned but not consistent pattern. He anchored a corner of the guest room at Tor House on a mass of bedrock, which he called Thuban in honor of the ancient polar star. He spanned one window with a single boulder, another with several square rocks, and over the main entrance to Hawk Tower he placed a perfectly shaped wedge as a keystone.

*Whole Earth Catalog* editor Stewart Brand wrote that Tor House "expressed more direct intelligence per square inch than any other house in America."[3] Jeffers expressed his connection to his house more poetically: "My fingers had the art to make stone love stone."[4] I know of no other person or building that better expresses the direct relationship between people and stone.

I returned to Tor House in April 2006 to meet Aaron Yoshinobu, a geology professor at Texas Tech University. For many years, he has been probing various Jeffers archive manuscripts and photographs, as well as "mapping" the stonemasonry and geology of Tor House and Hawk Tower. "One thing I like about Jeffers's work is that he talks about the importance of poets and artists creating things of permanence, of lasting value. You see that in the house and tower," said Yoshinobu.[5] "As a geologist, I relate to the intense connection and passion Jeffers found in rocks and mountains."

Jeffers wrote of rocks as the "bones of the old mother,"[6] the "world's cradle,"[7] and "old comrades."[8] Waves were "drunken quarrymen/ Climbing the cliff, hewing out more stone for me."[9] The surf "cheerfully pounds the worn granite drum."[10] During erosion the "hills dissolve and are liquidated."[11]

And it is clear Jeffers felt the tremor of at least one earthquake. He wrote:

*. . . the teeth of the fracture*
*Gnashed together, snapping on each other; the powers*
*of the earth drank*

*Their pang of unendurable release and the old resistances*
*Locked. The long coast was shaken like a leaf.*[12]

In a second, haunting description:

*The heads of the high redwoods down the deep canyon*
*Rippled, instantly earthquake shook the granite-boned ridge like a rat*
*In a dog's teeth; the house danced and bobbled, lightning flashed from the ground, the deep earth roared, yellow dust*
*Was seen rising in divers places and rock-slides*
*Roared in the gorges; then all things stilled and the earth stood quiet.*[13]

Yoshinobu's mother read Jeffers's work to him when he was a child, but it wasn't until he went to college and she gave him *The Selected Poetry of Robinson Jeffers* that he realized how much he liked Jeffers's language. Around the same time, when Yoshinobu was hopping between majors—first English, then astronomy—his mother also suggested he study geology. He thought it sounded boring but his girlfriend was taking a geology class and they went on a field trip together. He enjoyed it far more than she did, ended up majoring in geology, and went on to get a masters degree and doctorate, which he received from the University of Southern California in 1999.

His mother gave him another book of Jeffers poetry, *Cawdor and Other Poems*, during his work on his dissertation. The title poem, an epic narrative, centers on a family living at a ranch near Point Sur, thirty miles south of Yoshinobu's hometown of Pacific Grove. He read it every night in his tent in the Klamath Mountains. "There's something about growing up here and Jeffers's ability to grab the essence of this region. He made it universal," he said. "Taking Aeschylus and Euripides and recasting them in a new way on the Big Sur coast in a totally different rhythm. It was like nothing I had read before. It floored me."

Jeffers's passion, knowledge, and understanding of rocks and place centers on Tor House and Hawk Tower. "At the same time he's building these structures, he's working on his most evocative poems, that are not poems for this age but poems for ages on end," said Yoshinobu.

John Robinson Jeffers was born January 10, 1887, in Allegheny, Pennsylvania, to Annie Tuttle Jeffers, twenty-six years old and described as charming, outgoing, and musical. She was more than twenty years younger than his father, William Hamilton Jeffers, an intense and private man known to neighbors as "old Ichabod Crane."[14] A scholar of Greek, Latin, Aramaic, Babylonian, Hebrew, and Arabic, he taught biblical and ecclesiastical history at Western Theological Seminary, in Pittsburgh.

"When I was nine years old my father began to slap Latin into me, literally, with his hands," wrote Robinson, many years later.[15] To keep young Robin, as the family called him, better focused on his studies, Dr. Jeffers first planted a large hedge around their home, then moved to the country. The family also traveled extensively in Europe and lived abroad for many years. Robinson attended schools in Switzerland and Germany, often switching yearly. By age twelve, he could converse in Latin, read Greek, and spoke German and French fluently.[16]

His father's gift of two small books during the family's final year in Europe may have had as significant an impact as slapping and schooling. Robinson quickly abandoned one volume, the poems of Thomas Campbell, but he fell in love with the other, the poems of Dante Gabriel Rossetti. The florid verses intoxicated the impressionable fourteen-year-old, kindling a passion in Jeffers that led to subsequent encounters with Swinburne, Shelley, and Tennyson. He later wrote, "If I should ever wonder about the uses of poetry, I have only to remember that year's experience."[17]

The Jeffers family, which also included Robinson's younger brother, Hamilton, returned to Pittsburgh in 1902. His father placed Robinson in the University of Pittsburgh for one year before Dr. Jeffers's poor health forced them to move to Pasadena, California. Continuing his tradition of switching schools annually, Robinson now began again, this time as a sixteen-year-old junior at Occidental College. He attended classes in Greek, biblical literature, rhetoric, and astronomy. He wrote poetry for student magazines and in 1904 sold his first poem—"The Condor"—for twelve dollars.[18]

At Occidental, Robinson took two additional classes that had long-lasting effects on him, according to Yoshinobu. The first was a surveying class, which probably helped Jeffers when he began work on his stone structures in Carmel. He also studied geology using a standard text of the day, Joseph Le Conte's *Elements of Geology*. Le Conte, who was a pioneer member of the Sierra Club and an early proponent of evolution, compared the study of Earth with the study of the human body. Structural geology, defined as rocks and minerals and how they are arranged, was analogous to anatomy. The planet's physiology consisted of erosion and weathering, produced by water, wind, ice, plants, and animals. Historical geology, like human embryology, examined physical and biological changes over time. And like the human body, each facet of geology influenced the others.

"That's totally Jeffers. Le Conte has to be an early influence on Jeffers's notion of living, breathing rocks, the idea that everything around is part of one living entity," said Yoshinobu. "He can't wait to die to dissolve into calcium, which will form in the soil, which is part of the DNA of the first thing to crawl out of the ocean onto land. Oddly, Jeffers got his worst grade, an eighty-five, in geology."

Jeffers graduated in 1905 and that fall started at the University of Southern California. He planned on studying for a master's degree in letters, taking classes in oratory, Spanish, and advanced German. The latter course changed his life, for there he met the outgoing, intelligent, and beautiful wife of Edward "Teddie" Kuster, a prominent young lawyer. Una Call Kuster, two years older and three years married, had entered USC to escape the "incessant whirl of activity"[19] of her social life.

A hesitant romance began between the two but when Jeffers left again for Europe with his parents in 1906, he sent infrequent postcards. Upon his return, after taking a job translating German medical papers for his mother's physician, Jeffers entered medical school at USC. He also found time to swim and walk the beach with Una. By 1910 they knew they were in love but also knew they couldn't marry. Thinking that absence might help the heart grow colder, Jeffers moved to Seattle, with his parents and Hamilton, to attend the University of Washington School of Forestry. By late spring, Una and Robinson's romance appeared over.

He returned to Southern California in the summer of 1911 and ran into Una within an hour of arriving. Fate sealed their romance. In May 1912 Teddie sent Una to Europe to get her away from her young poet. With Una gone, Jeffers "drifted into mere drunken idleness."[20] He also produced his first book of poetry, *Flagons and Apples*. He paid for its publication with an inheritance from his uncle. The book's one review, in the December 8, 1912, *Los Angeles Times*, noted Jeffers's "distinctly novel and individual touch . . . [which] contains some of the best poetry I've seen in a dog's age—except of course, my own."[21] (If only modern writers could pen their own reviews, as Jeffers did.)

Una returned from Europe in November, prompted by Teddie's growing relationship with another woman. After Una and Teddie split up, she and Robinson began a more open life together.

Jeffers made it into the *L.A. Times* again on February 28, 1913. Under the banner headline "Love's Gentle Alchemy to Weld Broken Lives" the paper detailed the love triangle of Una, Robinson, and Teddie, calling it "a story so remarkable as to almost defy parallel." A day later a second feature on the "eternal triangle" provided more details, as well as a copy of Jeffers's poem "On the Cliff," with lines such as "our eyes were blind while my lips drank/Oblivious love at yours."

By this time, Robinson was back in Seattle. Una eventually followed him north to wait for her divorce from Teddie to be final. She and Robinson were married on August 2, the same day that Teddie remarried.

The Jeffers moved to Carmel in September 1914. They had not planned to. Their stated intention following their marriage was to settle in Lyme Regis, in Dorset, on the southern coast of England. By November 1913, however, Una was pregnant and they decided to stay in La Jolla, where they had settled. Una gave birth on May 5 to a ten-pound girl, Maeve. The baby died one day later. They then moved in with Robinson's parents, who lived in Pasadena, and planned to leave for England in the fall, but the outbreak of World War I ended that idea. A friend suggested they investigate Carmel.

They arrived by stagecoach from Monterey. "[When] we looked down through pines and sea-fogs on Carmel Bay, it was evident that we had come without knowing it to our inevitable place," wrote Jeffers.[22] They rented a small cottage. He wrote poetry and Una studied "certain aspects of late 18th century England."[23] They could see only three houses from the beach when they walked with their bulldog, Billie. Una described the time as "full and over-full of joy."[24]

Carmel is a spectacular place. It sits on a deep blue bay with a white sand beach. Rarely cold and rarely hot, the climate is a pleasing mixture of fog and not too many completely clear days. Outside of the central business district of either overpriced or too cheap tourist shops, the streets are quiet and without sidewalks. The older houses lack street numbers; mail only goes to the post office. And to the south, the foothills of the Santa Lucia Mountains drop treeless down to the famous Big Sur coast.

Carmel was both old and new when Robinson and Una arrived. Spanish navigator Sebastian Vizcaino chanced upon the harbor in 1602 and named it for his patron saint Our Lady of Mount Carmel. In 1771, Franciscan Friar Junipero Serra established a mission, made first of wood, then of adobe, and finally, in 1793, of sandstone, quarried one mile away. The mission thrived until 1834, when the Carmel priest moved to Monterey, the state capital. Eighteen years later the roof collapsed and the remains moldered.

In 1903 developer Frank Devendorf began promoting the property he had recently acquired around the bay. By the time the stagecoach dropped the Jeffers on the main road—dirt-covered Ocean Avenue—about 350 people lived in the village of Carmel. Most homes used kerosene for light, movies were a nickel, and news was posted on a bulletin board at the post office. Despite, or because of, the lack of amenities, Carmel had started to attract a well-known crowd, including Jack London, Upton Sinclair, Mary Austin, and Sinclair Lewis.

Two years after arriving in Carmel, Una gave birth to twin boys, Donnan and Garth. The family continued to rent a cottage and go on walks, including their favorite one, which wound along a grassy track and through acres of poppies and lupines to an open knoll topped by several granite boulders. Known as Carmel Point, the land had been a nine-hole golf course until World War I. In spring 1919 Robinson and Una bought two and a half acres (sixteen lots) on the point, which reminded them of barren knolls, or tors, they had seen in Dartmoor. (Never wealthy, they had a ten-thousand-dollar inheritance from Robinson's uncle and about two hundred dollars per month from a trust established when his father died in December 1914.) They found golf balls in their garden for many years.

Work on Tor House began in late spring at the property's high point, near the granite boulders known as the Standing Stones. Carmel contractor M. J. Murphy had submitted a thirteen-item estimate, including lumber, labor, cesspool, and profit, totaling $2,230. Because work progressed slowly, Jeffers apprenticed himself to the stonemason, a man named Pierson. "[Robinson] hadn't any skill of any kind so he did the hardest and plainest job (at $4.00 a day, I think)," wrote Una.[25]

The rocks came from the cove below the house. Murphy built a wooden track and used a horse to pull a cart up to the site. They also used the horse to help excavate the foundation, along with pick and shovel. Jeffers helped mix the mortar and place stones. By the end of the day, he was tired but happy, wrote Una. Murphy, Pierson, and Jeffers finished the house in August 1919.

To the Rock that will be a Cornerstone of the House

*Old garden of grayish and ochre lichen,*
*How long a time since the brown people who have vanished from here*
*Built fires beside you and nestled by you*
*Out of the ranging sea-wind? A hundred years, two hundred,*
*You have been dissevered from humanity*
*And only known the stubble squirrels and the headland rabbits,*
*Or the long-fetlocked plowhorses*
*Breaking the hilltop in December, sea-gulls following,*
*Screaming in the black furrow; no one*
*Touched you with love, the gray hawk and the red hawk touched you*
*Where my hand now lies. So I have brought you*
*Wine and white milk and honey for the hundred years of famine*
*And the hundred cold ages of sea-wind.*
*I did not dream the taste of wine could bind with granite,*
*Nor honey and milk please you; but sweetly*
*They mingle down the storm-worn cracks among the mosses,*
*Interpenetrating the silent*
*Wind-prints of ancient weathers long at peace, and the older*
*Scars of primal fire, and the stone*
*Endurance that is waiting millions of years to carry*
*A corner of the house, this also destined.*
*Lend me the stone strength of the past and I will lend you*
*The wings of the future, for I have them.*
*How dear you will be to me when I too grow old, old comrade.*

Yoshinobu and I entered Tor House through the narrow front door, on the lee side from the ocean. The compact twenty-foot-by-fifteen-foot living room felt somber due to a combination of dark, redwood paneling and cloudy light coming through the large west window and smaller south windows. Bookshelves ran along the base of the windows, as well as to the right of the front door, holding dark hardbacks with ragged-edged spines written by Yeats, George Moore, and Swinburne. Well-worn carpets covered the floor and absorbed any footsteps. In the far corner stood Una's Steinway piano. A few feet away, blackened by decades of smoke, was a tawny sandstone fireplace, a necessity in a house originally without electricity or gas.

Jeffers based his design on a Tudor barn Una had seen on a trip to England. Although he was over six feet tall, he built the house low as protection against the prevailing wind blowing off the bay and because the only heat came from fireplaces. Kerosene oil lamps provided light, and the Jefferses heated water in kettles in the fireplaces.

In a nook around the corner from the door was Una's built-in desk and portraits of Yeats and the twins, Donnan and Garth, along with drawings of Irish towers, a passion of hers. Above her, in a loft reached by steep stairs, Robinson would pace, pondering a line of poetry, periodically stopping to jot a finished thought. Family lore holds that during extended lulls Una would thump the loft with a broom and shout "Pace, Robin, pace." The family also slept in the loft, which is off limits to the public.

We passed through the doorway west of the fireplace into a small bedroom, where a black-and-white photograph of Una by Arnold Genthe is hung on the wall next to the bed. With her long dark hair wrapped around her head, she stares directly at the viewer with earnest eyes. No other photograph of Una reveals her piercing beauty as well. The bed was, as Robinson wrote in "The Bed by the Window," "unused unless by some guest in a twelvemonth." Jeffers died in this bed on January 20, 1962, during a rare Carmel snowstorm. In the same poem, he wrote, "I chose the bed down-stairs by the sea-window for a good death-bed."[26]

After leaving the bedroom we walked back through the living room into a corridor about five feet wide and twice as long. Originally the kitchen, books now lined its walls. It is hard to imagine how Una cooked for four in such a small space. In a red-velvet-wallpapered bathroom off the kitchen stands a red clawfoot bathtub. A leather shaving strop dangling by the little sink looked as if Jeffers had just used it.

As the Jeffers boys matured, Robinson decided to build a larger dining room off of the kitchen. The room felt more open and friendly than the rest of the house because of the two picture windows and lack of dark paneling, which also allowed closer inspection of Jeffers's beautiful stonework. Although people often thought of Jeffers as unfriendly and misanthropic—he famously had a sign posted on his gate reading No Visitors Until After 4 O'clock, by which time he and Una had headed out for their daily walk—Robinson and Una often entertained here. Friends such as Edna St. Vincent Millay, Charlie Chaplin, and George Gershwin sat around the six-foot-long oak table drinking homemade wine fermented from oranges, raisins, and rice. Entertainment included songs and performances, acted out in the Minstrel Gallery at the room's south end. They were warmed by the big corner fireplace, which Jeffers built on the exact spot where he found fire-charred stone five feet underground. ("The rock-cheeks have red fire-stains.")[27]

A locked door on the eastern wall leads into what had been the garage, the first building that Robinson constructed by himself. After finishing the garage, which has an arched opening, Jeffers read that he should have used bigger buttresses to support the arch. One of these expanded buttresses now jutted into the dining room. A second odd feature was the five-foot-tall door, which was at most two-thirds the width of a typical door. When Jeffers's son Donnan and his family later moved back to Carmel, they expanded Tor House and converted the garage into a kitchen, which required cutting an opening into the shared wall. Workers spent several hours trying to drill through the wall but made only a small hole. Two men came back with a pneumatic drill and needed two days to cut the hole. When one of them died of a heart attack after the second day, everyone decided to do no further work on the doorway.

Yoshinobu and I exited west outside through a normal-sized door to Thuban, the great rock Jeffers describes in "To the Rock that will be a Cornerstone of the House." Nearby was the group of rocks known as the Standing Stones, one of which the family called the Anvil Stone. When asked whether Jeffers mixed honey and milk on Thuban, Yoshinobu responded: "I think he really did it. Pouring wine and honey was another way for him to honor the stone."

Sitting out of the wind near the Anvil Stone, Yoshinobu expanded on some of the finer points of Jeffers's work on the house. Jeffers collected stones in the afternoon and at night, to avoid the scrutiny of curious Carmelites. Not all of the rock in the structures is local. Over the years the Jeffers family acquired material from their travels. Friends also sent them rocks. They include limestone from the Great Pyramid, lava from Vesuvius and Kilauea, stones from the homes of Lord Byron, George Moore, and William Yeats, pebbles from King Arthur's castle and the tomb of Cecil Rhodes, petrified wood and a meteorite from Arizona, and marble from Greece, Ireland, and Italy. (Jeffers did write once that he took stones from Carmel back to Ireland on trips; he did so to keep the balance of the world.) The collection makes me rather jealous.

Yoshinobu also noted how Jeffers's use of stones changed. In the main house, where he was only an apprentice, the pattern is straightforward. There is more regularity to the size of stones. "In the tower, though, there's a tendency toward big stones, things that look like they are enduring," said Yoshinobu. "The tower's just huge stones in all sorts of weird places."

From Thuban the path continues along the western side of the original house where the rich smells of rosemary and lavender mixed with a briny mist drifting in from the nearby ocean. The path ended at a small wooden gate in a stone wall, behind which opened a grassy yard. In the corner Jeffers built a pedestal for a sundial that Una had acquired in Cornwall in 1912. The still cloudy day prevented us from seeing if the sundial worked. Jeffers built both the wall, later raised for privacy, and pedestal in 1920, the same year he began work on Hawk Tower.

Tor House and Hawk Tower,
built in 1919–1925 by Robinson Jeffers, Carmel, California.

Una was the driving force behind the tower. She loved the ones she had seen in Ireland. The only extant sketch of a tower, drawn by Jeffers in early 1920, shows a simple round structure with a square door, two arched windows, and two porthole windows. "To make a round tower would have been redundant. He was done imitating. Hawk Tower is like nothing else," said Yoshinobu. "It's like taking all of those classical influences and bringing them to the furthest west that humanity is going to go. To the New World and then to the west coast of the New World. He could look further west to the eye of the world."

Jeffers worked with massive stones, some weighing up to four hundred pounds, to build Hawk Tower. The lower walls are up to six feet thick. On the first two levels he rolled the stones up a long inclined ramp. In one photo the narrow wood beam looks barely able to hold the weight of a single person. To reach the higher levels, Jeffers built a pulley and hoist system and lifted the boulders to one corner. He then rolled the stones along the highest wall to wherever he needed them.

Yoshinobu and I entered Hawk Tower through a doorway capped by a keystone chiseled with the letters *U R J.* Although Jeffers did not write in the tower, his desk and writing chair took up most of the room. In the low-angle light Yoshinobu pointed out a few letters and scribblings etched into the wood by the hard pencil Jeffers used to write his poems. One set of squiggles looked as though he crossed out a word or phrase. To the left a doorway led to stairs down into "the dungeon," an area built for the twins. "I am not sure if the arch in this doorway was supposed to be this way but it doesn't look well behaved. It doesn't look very pleasing," said Yoshinobu. "I always felt this room had a certain anger in it."

Back from the dungeon and in the main room, Yoshinobu closed the main door, revealing a "secret stairway" just wide enough for one person turned sideways to ascend. Jeffers built this passageway for the twins and modeled it on ones from English castles. Their stairways were designed so that a person had to enter left shoulder first, which allowed the climber to wield a sword in his or her stronger right hand, in case an attacker was following. Creeping up to the second floor required taking big steps and grabbing onto the granite walls; I was glad that I was followed only by a geologist. A door that blended into the room's paneling opened into Una's room with another fireplace, a small bed, an oak armchair, and a melodeon. This level also had an additional room, with an oriel window, where Una could sit and see the ocean and across the yard to where Jeffers was writing in the loft of Tor House. For a man with a reputation as a misanthrope, he seems to have been devoted at least to his family.

A steep exterior flight of stairs accessed level three and a small room. In another whimsical touch, Jeffers put two portholes in the west wall. Both came from ships that washed up on beaches near Carmel in the 1800s. (It is often reported that one of the portholes came from the ship Napoleon used to escape Elba. It didn't.)[28] The portholes are the "eyes" that I saw from the road when I first viewed Tor House and Hawk Tower. After crossing an inlaid, white marble floor and ascending a final set of steep stairs, we reached the top of the tower.

Ocean waves pounded against Jeffers's quarry of granite drums below. Cars passed by and passengers periodically looked up toward the house and tower. Stretching around the Jeffers property were houses with multi-car garages squeezed together like the sardines formerly canned in Monterey. One particularly ugly modern mansion lurked above Tor House, like a bully planning a hostile takeover. I don't think Jeffers would have liked how the new, oversized trophy homes intrude on his quiet property; he sold his once-virgin, then tree-planted, lots only to pay taxes.

When Jeffers completed the tower in September 1925 and before he planted his groves of cypress and eucalyptus, the land around was open and treeless. Una once described the view as extending south beyond the Carmel River to Point Lobos and the Santa Lucia Mountains, north to town and the Del Monte Forest, and east to the Carmel Valley. Jeffers's forest, along with the houses that later replaced many of the trees, now block the view.

Yoshinobu and I also looked down on Tor House and the remaining structures—the east wing, two additional garages, and a family room, which wraps around a courtyard. Donnan had helped build these additions, all of which are off limits to the public. Next to the window that had been the entrance to the garage grew a yew tree, under which are buried the ashes of Una and Robinson. In a marble slab on the tower, Jeffers inscribed Psalm 68: "Why leap ye, ye high hills? This is the hill which God desireth to dwell in."

The next day I walked along the coast to Jeffers's main quarry, the beach below his house. On a rounded, low wall of rock, I watched as the waves, Jeffers's "drunken quarrymen," struck the land. Water is an ideal stonemason. It weathers and erodes the rock, removing weak layers and leaving behind a sea-hardened building stone. Jeffers wrote so beautifully of the permanence of stone and yet here on the continent's end, his granite is continuously beaten, battered, and broken.

The weathered granite that Jeffers used for Tor House and Hawk Tower looks like the fog and low clouds that I associate with Carmel, although iron in areas has leached out and rusted, giving some stones an orange hue. When the sun does come out, the gray granite turns whitish to tan, with specks of shiny black mica twinkling in the sunlight. The plain, subdued color results from the rocks' most abundant mineral—plagioclase—the feldspar absent from the Quincy rock. A lack of hornblende and clearer, less smoky quartz crystals further make Jeffers's rock lighter colored than Willard's stone of choice.

Jeffers's building stones also contain alkali, or potassium, feldspar, the most abundant mineral in the Quincy rocks. Here, however, the mineral forms, tabular crystals, some up to four inches long, which indicates the magma cooled slowly and gave the crystals time to grow. They are clear to white and to some people look like the big-headed hobnails used to protect heavy boots or shoes. Being ignorant of the great lexicon of cobblers, I just think the crystals are distinctive looking. Geologists refer to this texture of large crystals set in a fine-grained groundmass as porphyritic. Fruitcakes exemplify this texture.

The story of Jeffers's granite began around 115 million years ago somewhere south of present-day Carmel. I use *somewhere* because a faction of geologists still debate the exact point of origin of Jeffers's granite.

Geologists agree that this southern birthplace of Jeffers's granite was at the boundary between the North American Plate and the Farallon Plate, which was covered by the Pacific Ocean. It was a region of geologic activity, with the Farallon advancing east and North America moving west. Like all oceanic plates, the Farallon was primarily iron-and-manganese-rich basalt with a thin coating of generally fine-grained sediments, which makes the oceanic plate, or crust, very dense, especially compared with a typical continent, which consists of lighter aluminum and silicon-rich rocks. When the two leviathans ran into each other, the dense oceanic crust began to slide under, or subduct, beneath the continental crust, a process that occurs today off Washington and Oregon, as well as in the Aleutian Islands, Japan, and the Andes.

Subduction zones produce three rock assemblages. The first is known as the accretionary wedge, basically all of the sediments that get scraped off the down-going plate, as well as a few slabs of basalt caught in the tectonic blender. Because the sediments are wet and squishy, they deform in a highly unpredictable way and produce chaotic folds and faults. The Franciscan melange of coastal California, best known as the star of John McPhee's *Assembling California*, resulted from scraping off Farallon's sediments. In subduction zones, these rocks occur on the continent side of the deep trench formed by the descending oceanic plate.

The second type of rock forms during the oceanic crust's prolonged dive beneath the continent. At depths of a hundred miles or so, in the partially molten layer known as the mantle, volatiles such as water and carbon dioxide began to bleed out of the oceanic crust. The water lowers the melting temperature of the surrounding rock, which starts to melt and begins to rise. When this magma reaches the surface it erupts through fissures and linked chains of volcanoes, known as arcs. When the volcanoes occur in an island, like they do in Sumatra or Japan, geologists call them island arcs. When they occur on the edge of a continent, such as the Cascades of my home state of Washington, they are continental arcs.

Not all of the magma, however, reaches the surface. These reservoirs of molten stone, known generically as plutons, solidify five to twenty-five miles underground. One of the best known and largest is the Sierra Nevada batholith. Like most subduction-generated rock, the Sierra consists primarily of granite, with varying amounts of chemically similar rock such as granodiorite, diorite, or tonalite, rock types collectively called granitoids. The volcanoes that erupted at the same time as the batholith cooled have long since eroded. Jeffers's building stones formed as a pluton in this manner, about 85 million years ago, and cooled about ten miles underground.

The third important component found in subduction systems are the sedimentary rocks that form in the forearc. The forearc develops in the region between the accretionary wedge and the arc. During subduction a variety of stresses pulls down on the crust in the forearc, generating a basin that fills with thick accumulations of sediments. The world's best example of an ancient forearc basin is the Central Valley of California, the four-hundred-mile-long lowland that runs from Bakersfield through Fresno and Sacramento to Redding.

In most subduction zones, if you travel from the ocean to the arc in order you would pass across the accretionary wedge, then the forearc basin, and finally reach the arc, usually covering a distance of about a hundred miles. This sequence can be seen in California traveling from the coastal Franciscan melange and its rich stew of dark sandstones and cherts through the black shale and tan sandstone of the Central Valley and up into the gray granite of the Sierra Nevada batholith. Geologists refer to this zone as the arc-trench gap, the distance from the folded and faulted ocean-derived sediments to the volcanoes, or the rocks that cooled under them.

"What makes Salinia really interesting is that it screws up that pattern of wedge-forearc-arc" said Dave Barbeau, a geologist at the University of South Carolina.[29] "We've got accretionary wedge rocks, the Franciscan melange, right next to or very close to arc rocks, the Cretaceous granitoids. Within five kilometers [three miles], some of the forearc rocks are also present. So the order is wrong and the distances are really wrong."

The term "Salinia" refers to an accumulation of rocks that makes up the central coastline of California. Geologically bounded on the east and west by the San Andreas and Sur-Nacimiento faults, respectively, and on the south by the Big Pine fault, Salinia comprises the rocks of the low mountain ranges (Santa Lucia, La Panza, and Gabilan ranges) that run north-south from about Santa Barbara inland of the coast to Santa Cruz. The rocks of Salinia include young sediments, older metamorphic rocks, and medium-age granites, which are Barbeau's Cretaceous granitoids and include the rocks used by Jeffers. Salinia has long troubled geologists, who have called it an "orphan," because for years no one knew exactly where the rocks originated.

In 2005 Barbeau coauthored a seminal paper addressing the origin of Salinia. His work was part of a long-term study based out of the University of Arizona, where he received his Ph.D. in 2003. The Arizona team hopes to put together a picture of plate tectonics in western North America. Barbeau's paper focused on two related models to explain the odd juxtaposition of Salinia and the surrounding rocks.

Model one proposed that Salinia formed between fifteen hundred and thirty-five hundred miles south, in southern Mexico, and then traveled north and wedged itself into the arc-trench rocks. Evidence for this long-distance movement came from what is known as paleomagnetism. When magma solidifies, iron-rich magnetite crystals within the molten rock align themselves relative to Earth's magnetic poles. By reading these minerals, geologists can determine the rock's latitude at the time of crystallization.

Barbeau and his co-workers favored a modified version of this model. Salinia moved north and injected itself into the arc-trench material, but instead of traveling thousands of miles, Salinia only moved about two hundred miles north, carried along by the San Andreas fault from around the Mojave desert.[30] Specifically, Barbeau hypothesized that Salinia once filled a gap between the Sierra Nevadas and the Peninsular Range, both made of granites of similar age and chemistry. The main line of evidence for the Mojave-Salinia connection came from the mineral zircon.

"You know there's the saying that diamonds are forever, but to geologists it's zircons that are forever," said Barbeau. "They are really resistant to heat and sedimentary processes. They basically never go away." The oldest known object on Earth is a 4.4-billion-year-old zircon crystal, about two human hairs wide, found in rocks from Australia. Zircon generally forms in igneous rocks, particularly in granites in subduction zones. When these granites weather and erode, wind, water, and/or ice redistribute the zircons and they end up in sedimentary rocks. By analyzing the various zircons in a sediment, geologists can reconstruct long periods of time, particularly because most sedimentary rocks contain zircons of diverse ages.

"Nearly all of the zircons we see in the Salinian sediments point to a North American origin as opposed to a southern Mexico one," said Barbeau. The zircon grains ranged in age from 80 million to 3 billion years old, with six periods of peak accumulation. Barbeau's research indicated that for five of the six peaks the western United States could be the only source of the zircons. Combined with other data, the zircons showed that after its initial formation Salinia remained in the south for about 40 million years until the San Andreas propelled the rock on its travels north.

In addition to the overwhelming zircon data, new research has shown that the original interpretation of paleomagnetism erred in the point of origin for Salinia. Only a handful of recalcitrant geologists still subscribe to the southern Mexico model. "This change in perception of how far Salinia has traveled is ideal for understanding the evolution of thought on accreted terranes," said Barbeau.

Like the Avalon terrane, which carried the Quincy Granite to the eastern edge of North America, Salinia is also a terrane, though with some differences. Avalon is an accreted terrane, meaning that it collided with its new home. It is also sometimes referred to as "allochthonous" or "exotic," which indicates Avalon traveled to reach its present spot. By definition an accreted terrane is also an allochthonous or exotic terrane. A suspect terrane, such as Salinia, is one whose origin is unclear. The terms are somewhat fluid and reflect personal choice more than rigid definition.

Geologists first began to focus on terranes after a landmark paper published in *Nature* on November 27, 1980. In "Cordilleran Suspect Terranes," Peter Coney, David Jones, and James Monger wrote that 70 percent of the land from the Rockies to the Pacific Ocean—the Cordillera—was a vast mosaic or collage of landmasses, which the geologists called "suspect terranes." More than fifty distinct terranes had collided with North America, including nearly all of Oregon, Washington, and Alaska, and most of California, Nevada, and Idaho. The paper did not indicate where the landmasses, which included island arcs, continental bits, and oceanic crust, originated, but hypothesized that scraps of land littered the Pacific from around 200 to 50 million years ago. The geologists' proposed mechanism of transport was eastward movement of the Pacific, Farallon, and Kula plates, as they progressively subducted North America.

The work of Coney and his partners was critical to the then two-decade-or-so-old theory of plate tectonics. It showed that not only did movement of big plates generate geologic features, but also that smaller slivers of land helped form a landscape. The terrane theory has been called "one of the most significant tectonic concepts" of recent years, particularly as it applies to the western United States.[31]

"When terranes first became popular, they were applied everywhere, probably too much," said Barbeau. Whenever someone couldn't figure out the exact story for an out-of-place rock it became a terrane. And the more exotic or suspect the better, which made the fifteen-hundred-to thirty-five-hundred-mile voyage of Salinia from southern Mexico seem possible. The "orphan" now had a parent. But over the years geologists have honed the terrane concept and learned not to apply it everywhere. As Barbeau observed, Salinia is a terrane but it has not traveled far and is neither exotic nor accreted.

"The only reason you have these granite rocks on the coast that Jeffers used for his house is because of this weird tectonic history," concluded

Barbeau. "Jeffers's house is that much more unusual because it's only that small slice of land between Carmel and Half Moon Bay where you would find these granitic rocks anywhere along the coast of the western United States."

To paraphrase Humphrey Bogart in *Casablanca*, of all the rocky knolls on all the West Coast Jeffers chose the one underlain by granite. And what a difference it made.

Before he started work on Tor House, Jeffers's poems lacked vitality. They are imitative, particularly of his early heroes Wordsworth and Shelley, often focus on love, and have simple, rhyming forms. But then he moved to Carmel and started his work with stone. The metaphors began to take shape. His education and local influences, his understanding of cosmological and atomic developments, his daily work with stone, each contributed to his poetry. "Jeffers always had this timeless, enduring imagery in his head and just hadn't found the voice nor the metaphor to make it sing. So in a sense Jeffers found granite and granite found Jeffers," said Aaron Yoshinobu. Jeffers's ideas crystallized in Carmel and gave him his distinctive voice.

Robinson Jeffers, 1930s.

Once his themes took shape, Jeffers kept working them throughout his life. Tragedy and suffering, the Big Sur coast, the beauty of Nature and its redemptive qualities. The poems are not always easy to read. His great epic verses, such as *Cawdor*, *Tamar*, *Roan Stallion*, and *Thurso's Landing*, are filled with incest, brutality, death, and pessimism and stretch for pages and pages, almost more novel than poetry. As one critic wrote, "Sometimes hard to stomach, they are always difficult to put down."[32] But many of his poems also contain short and beautiful passages describing his beloved landscape around Carmel.

His devotion to what he calls "my loved subject: Mountain and ocean, rock, water and beasts and trees"[33] is what makes his poetry resonate, particularly his shorter poems. He writes precisely and knowledgeably about landscape, both solid and oceanic, and its inhabitants. Seasonal change plays out in the mountains as they "vibrate from bronze to green/Bronze to green, year after year."[34] Gulls are the "slim yachts of the element."[35] Pelicans flying over Tor House "sculled their worn oars over the courtyard."[36] Cormorants "slip their long black bodies under the water and hunt like wolves."[37] Solomon's seal makes "intense islands of fragrance"[38] and eucalyptuses bend double "all in a row, praying north."[39] One can learn so much about the natural world from Jeffers's poetry that it is almost as if he has written a field guide to the Carmel coast.

But Jeffers's use of geology and geologic metaphors shines above all else. Weathering is "the endless ocean throwing his skirmish-lines against granite . . . / that fierce music has gone on for a thousand/Millions of years."[40] In an homage to evolution and geologic time, he wrote that "the wings torn with old storms remember / The cone that the oldest redwood dropped from, the tilting of continents, / The dinosaur's day."[41] During erosion "Cataracts of rock / Rain down the mountain from cliff to cliff and torment the stream-bed."[42] In contrast a resilient stone is "Earthquake-proved, and signatured / By ages of storms."[43]

Jeffers clearly paid attention to the natural world around him. Ever since his childhood he had had a connection to nature, but not until he settled in Carmel and worked on the land did he develop the knowledge and strength that gave him that passion to describe place. And this relationship centered on the house and tower he built from granite boulders on a low, barren knoll overlooking the sea.

"The place was maiden, no previous / Building, no neighbors, nothing but the elements, Rock, wind and sea," wrote Jeffers in a poem titled "The Last Conservative."[44] Tor House and Hawk Tower are the only structures he could have built for the site. How could I not love those buildings? In his ode to Tor House, Jeffers concluded, "My ghost you needn't look for; it is probably / Here, but a dark one, deep in the granite."[45]

# 4

# Deep Time in Minnesota—Minnesota Gneiss

*These masses bear very evident signs of a crystalline origin, but the process must have been a confused one.*
—William Keating, *Narrative of an expedition to the source of St. Peter's River, Lake Winnepeek, Lake of the Woods, &c. performed in the Year 1823*

NO MATTER WHERE you look at rocks, you are missing most of the story. Either erosion has removed layers, no layers were deposited in the first place, or the layers lay underground and cannot be seen. If you live in a city, the geologic story may barely equal a paragraph of time. For example, in Seattle, where I live, the oldest rocks you can see are the 55-million-year-old basalts of the Olympic Mountains. The skyline-defining volcanoes, such as Mt. Rainier to the south and Mt. Baker to the north, are each under a million years in age, and the last glaciers, which carved the modern topography, retreated only thirteen thousand years ago.

Going to a wilder locale may add only a few chapters to the story. In Moab, Utah, my home for nine years, you can pick up salt that had crystallized out of a sea 300 million years ago, climb a mountain that solidified as a hot hump of magma between sheets of sandstone 24 million years ago, or canoe between canyon walls carved by a river 2 million years ago. The story has expanded to a third of a billion years long, but many small gaps exist and no evidence occurs for the first 4 billion years of Earth's history.

Even the Grand Canyon, where you can see a mile of rock, fails to convey a complete story. The oldest layer, the Vishnu Schist, dates only to 2 billion years, and the youngest, Kaibab Formation, settled in a sea 245 million years ago. There is a known gap in the geologic record between 1.4 billion and 600 million years ago, as well as several slimmer pieces of missing time. I am not bothered by this. It is a fact of life.

Geologists have terms for such gaps of time. "Unconformity." "Disconformity." "Nonconformity." Each refers to a specific missing element. An unconformity is the most generic and refers to any break in the rock record. Again, either something was not deposited or something was removed. If that gap occurs between two layers of rock that are parallel, it is called a disconformity. A nonconformity refers to contact between two very different rock types, sort of like the relationship between a NASCAR mom and an NPR dad. An angular unconformity describes two rock units where there is an angle between the two layers.

One of the most famous of these unconformities is at Siccar Point, about thirty miles east of Edinburgh, Scotland.[1] There on a sunny June day in 1788, James Hutton sailed along the Berwickshire coast with his friends John Playfair, who later became Hutton's biographer, and Sir James Hall, who owned the boat. Three years earlier, at a presentation to the renowned Royal Society of Edinburgh, Hutton, an eccentric farmer and naturalist, had proposed that Earth was so old that no one could make a good estimate as to its age. Few had accepted Hutton's radical idea and the three men, or at least Hutton, were seeking evidence to prove his theory of an ancient Earth.

They began their trip on Hall's property at Dunglass Burn and proceeded east and south along the jagged coast, rounding several headlands before arriving at Siccar Point, where they found two rock layers. The lower strata consisted of gray beds of silt and sand, which stood on end like upright fingers of a hand. Resting directly atop the fingers were horizontal beds of red sandstone, which those in the eastern United States would call a brownstone.

From what we know about this adventure, Hutton gleefully explained to his friends that the gray stone, which he called schistus and which modern geologists call graywacke, had formed in the sea over vast epochs of time. Subsequent to its conversion to rock, the gray beds had been tilted vertically, and finally uplifted above the sea. More time passed, the sea rose, and sediments washed out of the nearby mountains and covered the schistus. Finally, Hutton told his fellow seekers of time that the gray and the red must have risen again, together as one. Clearly great ages of time had passed in order to generate such an unlikely wedding of rock. Our three

Scotsmen had no way of knowing how much time had passed between the schistus and the sandstone, but the beds at Siccar dramatically illustrated that Earth was far, far older than previously thought.

Geologists are still finding these gaps. In contrast to when Hutton found Siccar Point and could only recognize a relative gap, modern geologists can determine the absolute gap. They can take almost any rock, crush it to a powder, probe its chemical constituents on a microscopic level, and determine its age, whether in the thousands, millions, or billions of years.

Using these techniques, known as radiometric dating, geologists have discovered that a stone in the building trade has one of the longest story lines of any rock on the planet. The geologic tale of the Morton Gneiss (pronounced "nice"), quarried in Morton, Minnesota, stretches from its origin 3.5 billion years ago to 12,900 years ago. In that time the swirly pink, gray, and black, heavily metamorphosed Morton rocks were buried, baked, squeezed, warped, uplifted, injected with magma, and stripped bare by massive rivers. This 3.5-billion-year-long adventure created what one geologist calls "the most beautiful building stone in the country."[2]

My favorite Morton-covered building stands at the southeast corner of the main intersection in Morton, where State Route 19 crosses Main Street. It is architecture at its most utilitarian—rectangular, two stories tall, and brick. The builder did, however, incorporate some semblance of an aesthetic with the cornice and frieze, which have a pattern of outlined squares atop two horizontal rows of raised bricks resting on another row of inverted, stepped pyramids.

A faded red awning adds another touch of character, boldly proclaiming in large white letters, MORTON LIQUOR. Neon signs in the window hawk several brands of lite beers. Perhaps the store's owner thinks that Mortonians, or at least those who drink, should be concerned about their weight.

Below the cornice, polished slabs of Morton Gneiss cover the front of the building. Pink and black layers corkscrew around each other as if they are still fluid. Other layers look stretched and torn like taffy. Four-inch-wide eyes of black minerals, complete with white eyebrows, dot the variegated layers. This complex texture of light and dark bands typifies gneiss, a type of metamorphic rock formed from great heat and pressure.

Located two hours west of Minneapolis along the Minnesota River, Morton showcases a full complement of uses for its eponymous gneiss. The old high school features Morton blocks for accents, window lintels,

Morton Liquor in Morton, Minnesota.

and entryways. A few blocks away, the Zion Evangelical Lutheran Church, Wisconsin Synod, is made entirely of rounded, rough cut Morton stones, and the curbs on Main Street must be the most colorful as well as the oldest ones in the world at 3.5 billion years. In the cemetery above town, numerous headstones are made from Morton Gneiss. (Cemeteries are one of the best places to find the psychedelic rock; I have seen Morton headstones around the country.)

The first person to note the building potential of the Morton stone was Giacomo Constantino Beltrami. A former Italian soldier, he was a political exile who hoped to make a name for himself in America by finding the source of the Mississippi River. In June 1823 Beltrami had joined Major Stephen Long's expedition up the Minnesota River, then called the St. Peter's River and thought to be a possible source of the Mississippi. Beltrami wrote in his 1828 travel narrative that tears filled his eyes when he first saw the rock along the Minnesota River. "I should have given myself up to its sweet influence had I not been with people who had no idea of stopping for anything but a broken saddle."[3]

When he emerged out his reverie he noted that the "immense blocks of granite scattered here and there with such a picturesque negligence, might with small aid from the chisel be raised to rival the pyramids of

Memphis or Palmyra."[4] In lieu of grandiose structures, added Beltrami, the magnificent stone could easily be worked and fitted for barns, temples, or, for any local royalty, palaces.

Sixty-one years would pass before quarrying would begin in Morton, when Thomas Saulpaugh opened a quarry on a high dome of rock just west of town. By 1886 three hundred men worked the property. Other quarries opened on the southeast edge of town, along the railroad tracks, and by 1935 five companies operated holes, the biggest of which was owned by Cold Spring Granite Company. To move the stone, which quarrymen cut into Volkswagen bus–sized blocks weighing between fifty and eighty tons, Cold Spring built what was reportedly the largest swing boom in the world.[5] The steel mast rose 120 feet and supported a boom that could reach out 100 feet. Thirteen cables as thick as a man's wrist held the derrick in place, enabling quarrymen to move with ease blocks weighing up to a hundred tons.

Quarries sold the colorful stone under trade names such as Oriental, Tapestry, Variegated, and Rainbow granite. In what seems a rather ignoble fate, Cold Spring also sold crushed Morton as grit to go into turkey feed. At least the turkeys returned the Morton to the earth.

Records do not exist as to how much of it got shipped around the country or where it went, but some beautiful Morton buildings went up in the 1920s and 1930s. None were pyramids or palaces. The buildings include the tallest structure in Brooklyn, New York, the 512-foot-high Williamsburgh Savings Bank; the first modern planetarium in the western hemisphere, the twelve-sided Adler Planetarium in Chicago; and numerous structures now listed on the National Register of Historic Places, including Tulsa's Oklahoma Natural Gas building, Richmond, Virginia's, Old State Library, and Cincinnati's Cincinnati and Suburban Bell Telephone Company building.

Most of these buildings share two features. The first is that builders used the gneiss only at the ground level, because the Minnesota rock cost more but better resists urban agents of erosion, such as road salt, soot, and noxious vehicle emissions, than limestone and sandstone does. Second, the buildings were designed in the art deco, or moderne, style. The best Minnesota example is the Northwestern Bell Telephone Building in Minneapolis. Now owned by Qwest and formerly by AT&T, the twenty-six-story tower was built between 1930 and 1932.

The main entry of Morton Gneiss consists of three square recessed arches topped by a keystone shaped like a stylized bird reminiscent of

Egyptian art. Running along the top of the polished gneiss panels are classic moderne-style chevrons and zigzags against a background of vertical bands. Adjacent to the monumental entry rise geometric designs that resemble a tree with outstretched limbs dangling inverted triangles. A caduceuslike motif with lightning bolts replacing the traditional snakes surmounts each of the trees. Two adjacent doorways, each topped with another Egyptian-style bird, lead into a side entrance. Kasota limestone from southern Minnesota covers the remaining twenty-five stories.

Morton Gneiss was an ideal stone for art deco projects. It fit the style's aesthetic for unusual colors, particularly as a counterpoint to the light-colored, monochromatic stone often used above the base. (By highlighting the dark/light contrast, builders created their own unconformity. In the case of the Morton-Kasota contact, the missing time gap covering 3 billion years was what geologists call a "great unconformity.") Dark gneiss helped distinguish a building and set off the base from the surrounding structures. The Morton's swirled surface also provided a natural counterpoint to the era's prevailing fascination with machines and

Qwest Building, originally Northwestern Bell Telephone building, built 1930–1932, Minneapolis, Minnesota.

geometric patterns, as well as a complement to the fashion for abstract organic forms.

Although Beltrami recognized the potential of the Morton rocks, he was not a geologist. William Keating, also on Major Long's 1823 exploration, was the first geologist to see the Morton rock along the Minnesota River. "The character of these rocks was examined with care, and found very curious," he wrote in the official trip narrative. "It seemed as if four simple minerals, quartz, feldspar, mica, and amphibole, had united here to produce almost all the varieties of the combination which can arise from the association of two or more of these minerals."[6]

Keating encountered the Morton rocks at an interesting turning point, particularly on the subject of geologic time. For over a thousand years, biblical begats, begets, and begots had provided the means by which people had derived the age of Earth. By Keating's time, the best-known and most widely accepted day one of Earth was Sunday, October 23, 4004 BCE, according to Irish Archbishop James Ussher.[7] The biblical trap, however, was beginning to loosen in the early 1800s, because of the historic boat ride of James Hutton, John Playfair, and Sir James Hall.

In describing their momentous day, Playfair had written, "The palpable evidence presented to us, of one of the most extraordinary and important facts in the natural history of the earth, gave a reality and substance to those theoretical speculations, which, however probable, had never till now been directly authenticated by the testimony of the senses . . . What clearer evidence could we have had of the different formation of these rocks, and of the long interval which separated their formation, had we actually seen them emerging from the bosom of the deep? . . . The mind seemed to grow giddy by looking so far into the abyss of time."[8]

Geologists now know that Hutton, Playfair, and Hall were looking back 425 million years in time, to the age of deposition of the graywacke. The unconformity they saw represented a gap of 80 million years, which from a geologic point of view is a blink, but it was enough time for four-legged amphibians to have moved onto land, for forests to have spread widely across the globe, and for sharks to diversify and dominate the seas.

Hutton fleshed out the ideas the three men had discussed at Siccar Point in his book *Theory of the Earth; or an Investigation into the Laws Observable in the Composition, Dissolution, and Restoration of Land upon the Globe*. Playfair followed several years later in 1802 with *Illustrations of the Huttonian Theory of the Earth*. He had been forced to write his book, which covers the same material, because Hutton's dense and obtuse prose had prevented most readers from understanding his revolutionary geologic ideas.

As he noted in his title, Hutton based his ideas on what he could see out in the field. This in itself was a new beginning for most scientists, but he also observed that the planet was in a constant state of change and ever recycling itself. Dead organisms accumulated in the sea and made limestone. Mountains eroded to a "confused mass of stones, gravel, and sand," which consolidated into puddingstone. Rivers carried sands that became sandstones. Hutton's geologic reasoning has been summed up with a very Hillel-like statement, "The present is the key to the past."

One man who did penetrate Hutton's dense language was Charles Lyell, a twenty-seven-year-old lawyer and amateur geologist. In the year following Keating's visit to the Minnesota River valley, Lyell visited Siccar Point with James Hall. Lyell recognized the significance of the great unconformity between the schistus and the sandstone. In 1830 he wrote *Principles of Geology, Being an Attempt to Explain the Former Changes of the Earth's Surface, by Reference to Causes Now in Operation.*

*Principles* took Hutton's great statement and backed it up with numerous examples. Lyell also made two additional points: that natural laws did not change over time and that change occurred slowly and gradually and not catastrophically. Together these three principles form the core of uniformitarianism, one of the central tenets of geology. In addition, Lyell's book further pushed geologists to start pondering an Earth much older than the one described in the Bible. With *Principles* geologists now had their own scripture.

Although *Principles* pointed the way, acceptance of an ancient Earth came slowly. When Charles Darwin wrote in the first edition of *On the Origin of Species* (1859) that the valley of the Weald in southern England took 306,662,400 years to erode, critics lambasted him and he pulled the number in the third edition. Yet in 1862 one of Darwin's detractors, William Thomson—later Lord Kelvin—proposed an age for Earth as great as 400 million years.

Kelvin did not base his calculation on uniformitarianism but on the second law of thermodynamics, that a hot mass will cool over time. Since Earth had started out hot, as indicated by rocks such as the gneiss of the Minnesota River valley, it must be cooling. Few could dispute Kelvin's observation. As he gathered more data, however, he dropped his number to 100 million years and finally to 24 million years.

Not all geologists agreed with the physicist Kelvin. What about the slow, steady rates of geological and biological processes, which required more time than Kelvin proposed? By looking at sedimentary processes, such as deposition and erosion, geologists derived numbers between 3 million and 15 billion with the most popularly accepted age of Earth being around 100 million years.[9]

Compared to what we now know the age of the earth to be—4.5 billion years—a range of 24 to 100 million years seems to be puny. But consider the relative change between six thousand years and 100 million years. To make that leap requires four orders of magnitude compared with just one order of magnitude between 100 million and 4.5 billion. Hutton and Lyell had helped to push geologists across their biggest relative chasm of time. To get across the bigger, absolute chasm would require another radical shift in our understanding of the planet.

Geologists had long known that the Morton Gneiss was very old but not until 1956 when four University of Minnesota researchers published a crystallization date of 2.4 billion years did they learn how old.[10] The quartet based their age on a fundamental property of many elements, radioactivity, which simply means that an element is naturally unstable and constantly decaying. The best-known element for radiometric dating, as this technique is called, is carbon-14, which works for ages less than sixty thousand years. For ages of millions or billions of years, geologists turn to potassium, argon, lead, and, the most useful for the Morton, uranium.

When an element decays, or breaks down, it changes from an unstable isotope, called the parent, to a stable isotope, called the daughter, sort of like a 1960s flower child giving birth to an accountant. Isotopes refer to different forms of an element that have the same number of protons in their nucleus but a different number of neutrons. For uranium, which has several isotopes, the most important for geologists studying Morton Gneiss are uranium-238 (written $^{238}U$), which decays to form the lead isotope $^{206}Pb$, and $^{235}U$, which decays to $^{207}Pb$.[11]

What makes radioactive decay of uranium, and other elements in radiometric dating, useful to geologists is that the change from parent to daughter occurs at a measurable rate, called the half-life. The half-life of $^{238}U$ is 4.47 billion years, which means that 4.47 billion years after crystallization, half of the parent $^{238}U$ will have become the daughter, $^{206}Pb$. (Carbon-14 has a half-life of 5,700 years.) To figure out the age of formation of the uranium, and hence the age of formation of its surrounding rock, all geologists have to do is count the number of parent uranium isotopes and number of daughter lead isotopes.

Chemist Ernest Rutherford was the first to calculate the amount of the parent and daughter elements in a mineral. In 1904 he obtained an age of 497 million years. Three years later, an American chemist, Bertram Boltwood, showed that a uranium-bearing mineral (uraninite-$UO_2$) from Canada had formed 2.2 billion years ago.

Rutherford and Boltwood weren't geologists so radioactivity for them was more of a technical challenge than a way toward understanding geologic time. Geologists quickly recognized the insights to be gained with radiometric dating, and by the 1930s, it had become the accepted form of ascertaining the age of rocks. Researchers have continued to perfect their techniques both through improved technology and improved understanding of radioactivity. They have continued to refine the geological timescale, precisely dating eons, eras, periods, epochs, and ages. They have also continued to seek out and identify older and older rocks, including a beautiful building stone from Minnesota.

To analyze uranium in the Morton Gneiss, geologists studied zircon,[12] an extremely resistant mineral because of its high melting temperature and extreme hardness. In addition to zircon grains' long life, they have an unusual crystalline structure, which facilitates the incorporation of uranium from a magma, usually in concentrations of a few hundred parts per million.[13] (Zircon isn't radioactive because it contains so little uranium.)

Zircon's heat resistance also aids geologists in age dating. During deep burial or when intruded by magma, many minerals cannot withstand the heat these processes generate and end up melting. In contrast a crystal of zircon resists melting and instead attracts any uncrystallized zircon, and uranium, that might be in the introduced liquid. The new zircon forms a layer or rind on the original. Each subsequent high-temperature geologic event produces additional layers and each rind creates a time stamp as the uranium begins to decay, allowing geologists to date when that event occurred.

Adding new layers, like a tree adding rings, creates a technological problem. Although zircons occasionally grow to five-eighths of an inch long, the ones used to date the Morton Gneiss, and most other rocks, are about as wide as a two human hairs. The rinds are more miniscule, on the order of spider silk. When geologists first started to study the

Morton they did not have the technology to date separately the rinds and the core. Instead they analyzed the entire crystal and came up with what Pat Bickford, professor emeritus of petrology at Syracuse University, called "precise but inaccurate" ages for zircons.[14]

As their dating tools improved, geologists pushed the date of the Morton back. By 1963 it was 3.2 billion years old,[15] and in 1974 two researchers reported an age for the Morton of 3.8 billion years old, the oldest rock on Earth.[16] As one can imagine, there was much rejoicing. Fame was fleeting though. By 1980 the commonly accepted age of the Morton had dropped to 3.5 billion years. At that number, it was still the oldest, most commonly used building stone in the world, but the title of Earth's oldest rock belongs to gneiss found near Slave Lake in the Northwest Territories of Canada. Its age is 4.03 billion years.

In 2006 Pat Bickford led a team of researchers who obtained the Morton's most up-to-date age of 3,524 million years ago.[17] His date will probably be the one that sticks, mostly because he was able to take advantage of technology developed in the 1990s to analyze individually the core and rinds of a zircon crystal.

"If you want to know the answer badly enough, you will do tedious and persnickety work," said Bickford. To obtain a radiometric date, he starts by breaking rocks into plum-sized pieces, grinds them into powder with a jaw crusher, and separates out denser material ("like panning for gold"). He further isolates the zircons by floating the residue in a heavy liquid, which suspends the lighter particles, and by running the material through a magnetic separator. Once he obtains a good supply of zircons, he places them under a microscope and uses tweezers to handpick pure, crack-free crystals, each of which is half the size of a grain of salt. Finally, he mounts the zircons in epoxy and polishes them to half their original thickness. A typical puck of epoxy contains as many as a thousand zircons. Bickford does all this work at his lab in Syracuse before shipping the mounts out to Stanford.

There he uses the Sensitive High-Resolution Ion Microprobe, or SHRIMP, one of the tools that has enabled geologists to tease out the rind's geologic information. To date a zircon, the SHRIMP bombards the crystal's surface with a tightly focused beam of oxygen ions, which excavates micrometer-sized pits from the zircon crystal and strips uranium and lead atoms of some of their electrons. This gives the uranium and lead atoms, now called ions, an electrical charge.

Once ionized, the uranium and lead are accelerated through a strong magnetic field, which separates the ions on the basis of their mass. By changing the pull of the magnets, Bickford can precisely focus and measure specific uranium and lead isotopes. He usually counts $^{238}U$ and $^{206}Pb$, as well as $^{235}U$ and its daughter $^{207}Pb$, which have a half-life of 747 million years. The second set of isotopes provides an independent, corroborating age date.

"The SHRIMP has completely revolutionized geochronology. It means we can pinpoint specific regions of a zircon and get specific dates for specific events," said Bickford. With the SHRIMP uranium-lead method, Bickford can narrow down his dates to the point where the range of uncertainty is small enough that the number can be counted as accurate. For example, one zircon crystal from Morton Gneiss recorded three dates, which reflect initial formation and two subsequent geologic events. The average margin of error for each date is .005 of a percent.

Time is one of the hallmarks and central challenges of geology. How does one relate to billions and millions, or even tens of thousands, numbers not typically bandied about in daily conversation? To do so requires an openness and respect for the possibilities of time. For example, I can't see back in time but I can look at a sandstone and easily grasp that it was once sand that washed out of an eroding mountain chain, settled in a dune, disappeared under more sand, and finally got converted to rock.

I also trust the numbers used by geologists. The numbers are based on the laws of science and in the realms of observation, experimentation, and hypothesis. Scientists have tested and retested these numbers, not just on the same rocks or in the same laboratory or under the same conditions, but on different rocks, in different labs, and under different conditions around the world. The numbers have withstood the intellectual challenges of other scientists, scientists who had no interest, either financially, intellectually, or emotionally, in the truth of the numbers.

John McPhee coined the term "deep time" to describe the great abyss that made John Playfair so giddy. Deep time is what makes possible the almost imperceptible spreading of two plates to become the Atlantic Ocean. It is what allows a single finch on a small group of volcanic islands in the Pacific Ocean to evolve into thirteen species of finches, each adapted to a specific niche. Deep time is what helps with understanding the billions of years necessary for the formation of a rock unit in the Minnesota River valley that looks like a mixture of bubble gum and fudge.

According to our Morton amanuenses, Pat Bickford's zircons, the story of the Morton Gneiss began 3,524,000,000 years ago. Bickford found that the oldest zircons originated in the gray layers of the Morton. Because the gray layers have the chemistry of a type of rock known as tonalite, Bickford and many other geologists who have studied the Morton believe that the original source for the gneiss must have been a tonalite. Tonalites (named for rocks found near Tonale, Italy) are similar in composition to granite but a bit darker, thus in the beginning, the Morton was probably a drab, gray igneous rock with lighter speckles of biotite and chocolate chips of hornblende.

Bickford does not know where the Morton originated. He can, however, theorize about how it formed by looking at more modern tonalites. For example, 40 million years ago, during the collision of Africa with Europe, oceanic crust subducted continental rock and generated a magma that later solidified into the Italian tonalite. Colliding plates probably generated the Morton's tonalite, but geologists face a problem with making an absolute statement about the collision. Planet Earth 3.5 billion years ago did not look anything like the green and blue planet we now inhabit.

First, the color green probably didn't exist, or at least green life did not exist. The oldest fossil evidence for life are microbes found in 3.45-billion-year-old rock in western Australia, which means that when you see the Morton, you are seeing a rock that first cooled and solidified on a planet devoid of life. Second, what we think of as continents may not have existed either or they were a much less significant feature of our budding planet.

I use the word "may" because one of the great questions in geology is "When did plate tectonics begin?" At its most basic, plate tectonics describes the interaction of the dozen large and several smaller plates, consisting of continental or oceanic crust, which constitute Earth's outer layer. The word "tectonic" comes from the Latin *tectonicus*, pertaining to building. New crust forms at spreading centers in the oceans and moves away from its point of origin. Plates disappear when they dive beneath other plates at subduction zones but at the same time subduction generates continental crust. Although plate tectonics is one of the best known, most widely accepted, and easily explainable founding concepts of any field of science, a controversy centers on when plates began to form, interact, and disappear, or die, at least in the mode seen on modern Earth.

"All geologists were suckled on plate tectonics and they never accept that something other than plate tectonics existed," says Robert Stern of the University of Texas at Dallas.[18] He is one of the few who believes plate tectonics began as recently as only 1 billion years ago. Stern holds that early Earth was much hotter than it is today and crustal plates may have been too buoyant and too thick to subduct each other. If subduction was occurring on a young Earth, it should have generated three types of rock: ophiolites, blueschists, and ultrahigh-pressure (UHP) metamorphic terranes. Stern sees little evidence for these rocks prior to a billion years ago and without them, no plate tectonics.

Furthermore, Stern argues that a shift to an Earth dominated by plate tectonics must have had a profound affect on global environments. An increase in subduction-generated volcanism would have shot gases and fine particles into the atmosphere, cooling the planet and leading to a climatic condition that geologists call "snowball Earth," one of which froze the entire planet around 710 million years ago. "All in all, the arguments for early plate tectonics are fairly unconvincing," Stern concludes.

Kent Condie of New Mexico Tech counters that plate tectonics has operated since at least 2.7 billion years ago and possibly as far back as 4.4 billion years, the age of the oldest known mineral, a zircon from Australia. He challenges Stern's concerns about lack of subduction-related rocks by observing that a hotter young Earth altered how plates subducted, and that resulted in fewer ophiolites, blueschists, and UHPs, although all three rock types do exist as far back as perhaps 3.8 billion years ago. Condie also cites rocks such as the Morton's original tonalite, which are widespread in the Archean era (3.8 to 2.5 billion years ago) and require oceanic crust and subduction for formation. "It may be that plate tectonics did not begin globally all at once," says Condie. "It may have been episodic but by 2.1 billion years it was continuous."[19]

Condie makes an additional point. He observes that no other planets in our solar system have higher life forms, mostly because plate movement fostered our oxygen-bearing atmosphere by creating continents. Continents allowed for the evolution of photosynthetic plants, the source of oxygen. Furthermore, the shallow marine shelves that develop on the edges of continents are an ideal place for carbon dioxide to be deposited, in the form of calcite. If such deposition did not occur, carbon dioxide would accumulate to dangerously high levels in our atmosphere. "Without plate tectonics, humans wouldn't exist," says Condie.

No matter when and where the Morton formed, its birth as an igneous rock raises another question. What was the source of the magma?

Magma doesn't just appear magically, it has to have a parent rock that melts to form a liquid. The process begins in the asthenosphere, the partially molten layer of the mantle made primarily of peridotite, a greenish rock rich in the mineral olivine. The asthenosphere starts about 45 miles below the surface and extends to about 150 miles deep.

When asthenospheric peridotite partially melts, it produces a magma that solidifies to basalt, similar to what erupts in Hawaii. A good analogy is the partial melting of a Popsicle as you eat it outside on a hot day. As you hold your frozen treat, it invariably melts, sending a stream of sugary syrup down your arm. When you take the next bite of your icy confection, you will notice it has a slightly different taste and texture. The Popsicle is still solid but less sweet and more granular with icy particles. When a peridotite partially melts, it loses various elements such as iron, magnesium, and calcium to magma, but still remains a peridotite.

In the geologic process, less than 3 percent of the total peridotite changes to a basaltic magma and the rest remains as peridotite. The most common place to find basalt is at spreading centers where plates begin to pull apart from each other. Most of these spots are what geologists call a midocean ridge, such as occurs in the middle of the Atlantic Ocean or off the coast of Washington State where the Juan de Fuca Plate moves away from the Pacific Plate toward the North American Plate.

When basalt partially melts, the resulting magma can generate a tonalite. This process occurs most often at a subduction zone, where the basalt dives into the planet, warming roughly twenty-four to thirty-two degrees Celsius for every mile it descends. When it reaches temperatures of seven hundred to eight hundred degrees Celsius it partially melts. Because the melt is only partial, on the order of 10 to 15 percent, it is not unusual to find basalt inclusions in a tonalite, sort of as if chunks of Popsicle also dropped onto your arm at the same time the Popsicle melted.

In the Morton Gneiss, basalt appears as solid black rafts floating in the gray and pink swirls. The rafts range in size from a few inches to sixty-five feet across. Geologists refer to these anomalous blocks as enclaves or xenoliths, meaning foreign rock. Mark Gross, who has worked the Morton quarries for Cold Spring Granite for twenty-eight years, called the smaller of these variously shaped blobs "knots" or "cigars" (pronounced *see*-gars).[20] Quarrymen don't like the knots and cigars because buyers don't want such imperfections marring their stone.

Unfortunately it is impossible to date the Morton's basalt rafts because they lack zircons. Bickford believes there is a strong possibility that the basalt originated more that 3,524 million years ago but the rafts could also have formed during a later geologic event, which melted the basalt only enough to generate more basalt that injected itself as dikes into the surrounding rock. Such dikes are not unusual in other rocks that formed from basalt.

After recording the first crystallization of the Morton tonalite, the zircons logged geologic events at 3.42, 3.385, 3.14, and 3.08 billion years ago. Little field evidence for these events remains. The only feature that records one of them may be veins of light-colored rock peppered with black minerals that shoot randomly across the rock. Gross called the veins "big, ugly ropes," and like knots they lower the value of the stone.

The zircons went haywire 2,680 million years ago. Not only did the Morton rocks get folded like a gymnast, but they also got an infusion of pink magma that finally, after nearly a billion years of boring gray, gave the rocks their distinctive coloring. The fact that several nearby rock units record this same date means that an epic collision altered the Morton. That collision was between the block of rock that was the Morton's home, known as the Minnesota River Valley terrane, and a much larger mass of land, known as the Superior Province.

Mark Schmitz, a geologist at Boise State University who also studies the Morton, compared that impact to another impact that occurred only 60 million years ago. "Minnesota at 2.6 billion years ago would have looked like the modern Himalayas," he said. "There would have been a space problem as the two blocks came together. They would start to deform and thicken. As one block slid under the other, pressure and temperature would rise and the rock would start to melt." Or as another pair of geologists wrote, the collision resulted in "manifestations of constipated subduction."[21]

Once again, partial melting would occur, with the tonalite spawning a melt but this time the melt was pink; pink magma that injected itself into the gray tonalite as veins and pools. Heat didn't just melt rocks, it also weakened the Morton to a taffylike consistency and caused it to swirl, surge, seethe, and eddy. Most of the rock was still solid but the mountainous pile of material above was squeezing and deforming the Morton like toothpaste and metamorphosing it into a gneiss. Similar changes may have occurred at the earlier events recorded by zircons but the continental collision 2,680 million years ago erased previous textures and generated a metamorphic rock consisting of the bands of dark and light minerals that characterize gneiss.

The rock did not respond homogenously. The basalt rafts acted plastically, and either bounced back to their original shape or broke into fragments. The pink granite and tonalite acted like Silly Putty, deforming but not breaking under pressure; but the tonalite was less fluid than the pink granite.

One of the beautiful aspects of the Morton rocks is that you can see this give and take of rock. In one panel pink dominates, in another gray, and in a third, rafts of jet black basalt sit like islands awash in a sea of pink. Some Morton building panels look like still photographs of streams of blood flowing through arteries, a texture that quarry workers call veiny. But the dominant pattern resembles a series of pictures taken while stirring together cans of pink and gray paint. Quarrymen call this texture flurry.[22]

No matter which texture one sees in the Morton, the rock seems to be constantly in motion. Nothing is static. Although the rock records events that took place between 3.5 and 2.6 billion years ago, it is the most alive rock I have ever seen.

The Morton Gneiss story, however, did not end with collision 2.6 billion years ago. Two billion years ago, plate tectonic action thrust the Morton up and for the first time in its multibillion-year existence, the pink and gray rock was at or very near the surface of the planet. After 2.4 billion years of action, the excitement ended. Its story, though, was not over yet. One more significant geologic event would have to hit the Morton in order for it to be exposed at the surface, but that landscape-altering process would not occur until just thirteen thousand years ago.

Cold Spring Granite still owns and operates the Morton quarry, although they dismantled and scrapped their record-sized boom derrick in 1996. As happens with every product, fashion waxes and wanes and the Morton has been in a long-term wane; Cold Spring quarries stone at Morton for only a few months of the year. When I inquired about visiting the quarry, Dan Rea, vice president for Cold Spring's Commercial Division, was kind enough to open the site for me.

My guide was Mark Gross. We reached the quarry by driving through a gate with a No Trespassing sign south of town, down a dirt road, and parking next to an abandoned metal shed. Like every other quarry I have seen, the multilevel, football-field-sized stoneyard was strewn with massive blocks of cut stone and rubble piles of cut and broken stone. It also had the requisite scary-looking pit of cloudy water, which usually designates the oldest, now unused, portion of a quarry. Rusty water streaks stained the older cut walls, which towered fifty to sixty feet above the quarry's main floor and provided good nesting locations for swallows. Other walls were dotted with hundreds of parallel grooves as if troupes of industrious clams had been in a synchronized burrowing competition. These holes had been drilled to break blocks off a quarry wall.

"We now only have a few men work the quarry and they are specialists," said Gross. Cold Spring removed their big derrick because trucks and front-end loaders are more efficient. Working a boom required at least four men—one to operate the hoist, one to signal the operator, and two to attach the derrick cable to the block. Operating the newer machines requires only one man and is much less dangerous since multi-ton blocks no longer dangle from steel cables.

Gross called the process of quarrying "building a loaf." Imagine a squared off quarry wall, flat on the top with two perpendicular, vertical faces, one trending north, the other west, forming a corner. First, a quarryman uses a hydraulic drilling machine, which both pounds and spins a carbide-tipped drill, to cut a horizontal tunnel, up to eighty feet long, into the base of the north face. The three-inch-wide hole runs parallel to and eighteen feet from the base of the west face. In step two, the driller stands on the flat top and pierces the end of the horizontal hole with a vertical shaft as long as eighteen feet. To cut the rock, he threads a diamond-impregnated wire into the vertical shaft and down the horizontal tunnel and makes a loop by reconnecting the wire's two ends. The loop feeds through a machine that moves the wire like a conveyor belt. As the wire slowly cuts into the rock, the machine tightens the loop and it slices through the rock. This first cut of the loaf takes about fifty-eight hours in the Morton rocks; cutting softer granite takes half the time.

The quarrymen have two options at this point. They can either repeat the drilling and slicing process every six feet and fashion slabs eighteen feet high by six feet wide by eighty feet long or they can make one shorter cut and create a loaf eighteen feet high by eighteen feet wide by eighty feet long. Since both the slabs and loaf are still attached at the base, the driller drills a series of horizontal holes seven inches apart at ground level back to the eighty-foot-long horizontal hole he drilled. He then pushes sticks of Dynashear, roughly equivalent in force to about one-twentieth of a stick of dynamite, back into the holes, and detonates (or shoots) a slab or loaf, which pops free.

A buyer's need and stone quality dictates whether the quarryman cuts a slab or a loaf. Mausoleums may require wall-sized panels of rock and use of a loaf, which the driller drills into slices, six feet thick by eighteen feet wide by eighteen feet tall. In contrast, cladding or countertops use a long, skinny slab. To work on the slab, drillers drive wedges into the long gap made by the wire saw and force the slab to tip over onto old tires the size of a car or onto a pile of broken-up stone. Smaller blocks are made by drilling, which creates the burrowlike channels I saw on some walls of the quarry.

Unlike sedimentary rock, gneiss does not have to season or cure. It can be worked immediately. No further cutting, however, occurs in Morton. Blocks and slices get moved via a front-end loader onto trucks and transported to the Cold Spring Granite factory, in Cold Spring, Minnesota, about ninety miles north. Again, Dan Rea set up a tour for me.

After putting on a yellow hard hat and clear plastic goggles, I entered the cavernous fabrication, or milling, plant. We started at the gang shot saw, which looked like a bread slicer on steroids. Instead of sharp blades, however, the machine used rows of parallel steel plates that cut through the stone by moving back and forth, like a reciprocating saw. The flat, quarter-inch-thick, two-inch-tall steel plates, each about fifteen feet long, don't actually cut the stone but grind a slurry of water and steel shot—broken up bits of steel about half the size of a grain of rice—that do the cutting. Operating twenty-four hours a day, the incredibly noisy gang saw, named for its gang of blades, cuts through an eight-foot-thick block in three to four days. It can cut panels as thin as an inch.

Cut panels, which moved through the building via bright yellow overhead cranes, next received a surface finish. The first finishing machine used pie-pan-sized, diamond-encrusted buffers and could produce a finish ranging from glassy smooth to coarse and nonreflective. For a rougher finish, Cold Spring had a machine that resembled a pizza oven and sounded like a jet. Known as a thermal finisher, its eighteen-hundred-degree Celsius torch expanded and exploded surficial feldspar crystals, leaving a textured surface that works well for paving stones. Cold Spring formed rough, natural-looking faces with a stone splitter, which worked like a slow-motion guillotine to crack open a rock. When the stone cracked it sounded like a gunshot.

Cutting took place on the other side of the building. Computer-controlled, diamond-tipped circular saws cut most finished panels. Cold Spring had several with a variety of blades. They use one of the blades mounted on a overhead arm to make round columns by running the blade down the length of a column, rotating the column slightly, and moving the blade slightly out or in and down. Smoothing out the rough edges requires hand grinding and polishing. Humans also perform some of the most high precision work, such as cutting floral patterns into the rock.

For other intricate work, a high-pressure water jet can accurately cut to within one one-hundredth of an inch. The water shoots out at over nine hundred miles per hour at pressures of up to sixty thousand pounds per square inch (psi) and can cut stone up to four inches thick. By comparison, a typical fire hose operates at between one hundred and three hundred psi and a household faucet at between sixty and eighty psi.

Cold Spring primarily sells the Morton Gneiss for buildings, monuments, tombstones, and mausoleums. It is not popular, selling about 8,000 cubic feet annually, compared with Cold Spring's best sellers—a gray granite from Minnesota and a speckled, red-and-black granite from South Dakota—both of which sell more than 120,000 cubic feet per year. Despite the low sales for the Morton, Cold Spring vice president Dan Rea was optimistic. "Trends change. There is always hope."[23]

Twelve years after William Keating and Giacomo Beltrami ascended the Minnesota River, London-born George William Featherstonhaugh (pronounced Fanshaw) traveled up the waterway in a birch-bark canoe that carried eight men and thirty-five hundred pounds of supplies. Known to the native people as the Stone Doctor and to the federal government as Geologist to the United States, he reached the valley of the Minnay Sotor, as he spelled it, in September 1835. Featherstonhaugh noted the unusual gneiss, collected specimens, and made a perceptive observation: "It is evident that in ancient times . . . the volume of the river was many times greater than it is now."[24] He would be astounded by how much water once flowed down the Minnesota River valley.

This giant predecessor to the Minnesota River is called the River Warren.[25] When it flooded thirteen thousand years ago, Warren shot down the valley and carried as much as 100,000 times more water than what Featherstonhaugh encountered. Fed by runoff from one of the greatest lakes to cover the planet—Lake Agassiz—Warren spread across the entire valley, purging the land of soil and plants. The turbulent water also stripped away cobbles, rocks, and boulders deposited by previous glaciations, revealing the wonderful pink, gray, and black Morton Gneiss that lay below.

This final chapter of the Morton Gneiss's 3.5-billion-year-long story began during the last ice age, when a tongue of the Laurentide Ice Sheet retreated north back into Canada. At its maximum the several-thousand-foot-thick glacier had covered most of our northern neighbor and extended as far south as Iowa, but by 13,400 years ago, the climate had begun to warm and the ice was disappearing. Lake Agassiz formed along the southern boundary of the ice, at about modern-day Winnipeg. It spread mostly west to east, but one arm pushed south down into the United States along the modern-day Minnesota/North Dakota border.

Named for the great naturalist Louis Agassiz, Lake Agassiz lasted for over five thousand years and eventually grew to cover 580,000 square miles, about the size of Alaska.[26] Four outlets drained the lake, which reached a maximum depth of 2,500 feet. One drainage carried water north to the Arctic Ocean, one east out the St. Lawrence valley to the Atlantic, one northeast through Hudson Bay, and one south to what later became known as the Mississippi River valley—the River Warren. They did not drain at the same time.

For 99 percent of its life, Warren flowed at an average of 1.6 million cubic feet per second, or about three times the flow of the modern-day Mississippi River at its mouth and thirteen hundred times the flow of the Minnesota River near Morton. But then some environmental change triggered a flood of epic proportions. For days and maybe weeks, the river ripped at up to sixty miles per hour down the valley, a grayish-brown, roaring, churning soup of mud, debris, plants, boulders, and animals. Total estimated volume would have about equaled the combined average annual flow of the ten largest rivers on Earth.

Geologists don't know how many of these megafloods scoured the Minnesota River valley. The last one hit 12,900 years ago and all of Lake Agassiz drained in a cataclysmic flood 8,400 years ago. Geologists do know, however, that without the River Warren floods and the river's constant, high-volume flow out of Lake Agassiz, the Morton Gneiss may never have been revealed. Instead, tens to hundreds of feet of glacially deposited sediments would still be covering the rock and one of the planet's longest geologic stories could not be told.

I find it compelling that the last chapter of the Morton story is also the final global geologic event to hit the planet. From deep time to modern time, the Morton Gneiss has been present. It existed when life evolved on Earth, when modern-style plate tectonics began to operate, and survived the ice age. As one geologist says of the Morton, "It's got it all." Plus it is a damned gorgeous rock.

# 5

# The Clam That Changed the World—Florida Coquina

*And under Anastasia's verdant sky,*
*I saw St. Mark's grim bastions, piles of stone.*
*Planting their deep foundations in the sea,*
*And speaking to the eye a thousand things,*
*Of Spain, a thousand heavy histories.*
—Ralph Waldo Emerson, "St. Augustine"

HARDLY ANYONE THINKS that clams changed the world. Most are benign bivalves toiling away in the sand or resting quietly in the sea. If we do consider them, we are usually thinking about food, though perhaps the most famous mollusk in the world is the one that supports Venus in Botticelli's *The Birth of Venus*. Botticelli depicts the Roman goddess of beauty and love using the shell as a mode of transport as she arrives on land, blown there by the winds. Although unrealistic as a way to travel, Botticelli's shell does fit the classic image of a clam, something trod underfoot.

Thus some may find it odd that a bivalve, and one much smaller than the one that carried the lovely Venus, was seminal to the early colonization history of North America. Carolina governor James Moore was the first to discover the power of the clam when he lay siege to the Spanish colonial town of St. Augustine, in 1702.

Ambitious and greedy, Moore had arrived in Charles Town (later shortened to Charleston) in 1700 to govern the southernmost of England's American colonies. Founded in 1670, Charles Town had suffered smallpox, an earthquake, fire, and the yellow fever, but was a thriving town of four thousand when Moore arrived.[1] He recognized, however, that in order for Carolina to survive and prosper, the English had to defeat Spain's strongest North American colonial outpost, St. Augustine, in what is now Florida. In addition, he worried about the growing strength of France, which had established Louisiana in 1699 and had recently allied itself with Spain.

A French-Spanish partnership had developed because of the death of Spanish monarch Charles II in November 1700. The childless Charles had named his great-nephew Philip of Anjou as his successor. Philip was also Louis XIV's grandson, and Philip's ascension would unite Spain and France and pose a significant threat to England. To counter Philip's claim to the Spanish throne, England's William III allied with Holy Roman Emperor Leopold I to support Leopold's son as heir to the Spanish crown.

At stake was domination of Europe and the New World. Spain controlled Mexico, southwestern North America, Central America, and Cuba. France laid claim to North America from the mouth of the Mississippi River north to Canada, and England dominated the eastern seaboard, Jamaica, and Barbados. In May 1702 fighting began in what is known in Europe as the War of the Spanish Succession and in America as Queen Anne's War, after the queen who assumed the English throne following William's death.

When Moore learned of the fighting in August 1702 he recognized its significance and suggested to his Commons House of Assembly that "the takeing of St. Augustine before it be Strengthened with French forses opens to us an easie and plaine way to Remove the French (a no less dangerous Enemy in time of Peace than Warr)."[2] The Assembly approved Moore's plans and provided two thousand pounds sterling for expenses. His force of five hundred English and three hundred Indians sailed south from Charles Town in fourteen boats under the command of Moore and Colonel Robert Daniel.[3]

About fifty miles north of St. Augustine, the Carolina army attacked their first Spanish outpost, a guardhouse and small village at the north end of Santa Maria Island, at midnight on November 3, 1702. Two additional villages on Spain's northernmost Atlantic Coast settlement succumbed the next day. Moving south, by November 5 the English had destroyed the missions of San Juan del Puerto, Santa Cruz, and Piribiriba. Nothing now stood between eight hundred militia and St. Augustine. Three days later Moore reached St. Augustine with thirteen ships and many of his men. Daniel had taken the rest of the soldiers south overland.

Two hundred and forty nine Spanish soldiers under the command of Governor Joseph de Zúñiga y Cerda defended St. Augustine. Zúñiga had recently told the town's inhabitants of Moore's successes to the north, warned them of attack from the sea and from Daniel's land force, and finally ordered everyone inside the town's fort. He did not inform them he had sent urgent pleas for help to Havana because St. Augustine was short of men and ammunition. Nor did he tell them that the infantry consisted partly of old men, invalids, and young boys and that the gunners "had no service record, lack discipline, and have only a slight knowledge of the [fort's] bronze and iron guns." With everyone crowded into the fort the townspeople probably surmised the situation.

Despite the lack of manpower and firepower, and a fort overcrowded with fifteen hundred people and their farm animals, Zúñiga's situation was not completely desperate. Well located, St. Augustine stretched for about one-half mile south of the fort along the Matanzas River. A marsh protected the land north of the fort. East of the river a barrier island, Anastasia Island, reduced coastal access. At Anastasia's north end, hard-to-navigate sandbars created a treacherous entrance across the narrow inlet to St. Augustine's harbor. The west offered the only easy route to town.

Zúñiga's fort also provided a significant advantage against invaders. Known as the Castillo de San Marcos, its construction had been completed only six years earlier, and it was the tenth fort in St. Augustine. The previous nine had succumbed to pirates, fire, and water. The castillo's layout consisted of an open, 150-foot-wide square courtyard surrounded by storage rooms, living quarters, a jail, and a chapel. Arrowhead-shaped bastions with 90-foot-long sides jutted out from each corner. The white plastered stone walls were 16 feet thick at the base and rose 26 feet above a moat, which could be flooded with seawater. When the fort's gates closed, the castillo was a secure island with three freshwater wells and enough food to last several months.

What made the fort nearly perfect for Zúñiga's situation was the stone used in the walls. Quarried from Anastasia Island and known as coquina (ko-*kee*-na), it looked like what you'd get if you took a mound of whole and broken shells, mixed in a dilute solution of Elmer's glue, and let it dry. People liken coquina's consistency to a Rice Krispies Treat or a granola bar with shells and shell fragments replacing the oats. Either way, the dominant component by far of coquina was a shell, from the clam that changed colonial history. These shells gave coquina a property found in

Aerial view of Castillo de San Marcos, St. Augustine, Florida.

no other rock. Instead of breaking or cracking when hit by cannon shells, the cavity-rich coquina absorbed or deflected the iron projectiles.

Zúñiga did not know of the unique qualities of coquina. No one did. No one had yet attacked the fort. And no other significant building in the world had been made of coquina.

Moore began his siege on November 10 when Colonel Daniel arrived with his men after marching down the coast from Piribiriba. With the Spanish holed up in the castillo, the English took all of St. Augustine, but not before the Spaniards stampeded 160 cattle through Daniel's men and into the dry moat. The English set up around the perimeter of the castillo and began to dig trenches to get closer. They also sailed their boats across the sandbars that protected the harbor and began to fire cannons from sea and land. The Spanish responded by burning strategically located houses where Moore's men could hide and shoot anyone entering or leaving the fort.

As the siege progressed, Moore's circle tightened. By November 24, he had located four of his biggest guns only 750 feet from the castillo. At one point, a twenty-four-hour-long battle broke out with the British firing canisters, round shot, bar shot, and broken glass. The cannon fire could not break the walls of clamshell.

Moore's men continued to dig their trenches closer to the fort, supplemented by erecting rows of gabions, rock-filled cages that provided a protected shooting site for gunners. They also burned the southern end of St. Augustine. By December 19 they had advanced to within pistol shot of the castillo, and still their artillery did little damage to the massive, spongy walls. They would get no closer.

Four Spanish man-of-war gunships arrived the day after Christmas with supplies and men from Cuba. Moore realized the hopelessness of his situation, burned four of his ships, abandoned four others, set fire to the remaining houses in St. Augustine, and retreated north on foot to Charles Town.

When the gates of the Castillo de San Marcos reopened on December 30, the Spaniards found little left of St. Augustine. The English had destroyed the main church, the governor's palace, and all farms, fields, crops, and cattle. Only twenty houses remained in the desolate landscape. But St. Augustine was still Spanish, and Spain still retained control over Florida and its lucrative trade routes. The English would not attempt another attack on St. Augustine for thirty-eight years. All that had saved the Spanish was their castillo made of clamshells.

Joe Brehm has been a National Park Service ranger at the Castillo de San Marcos for thirteen years. Usually he wears the typical green and gray NPS uniform, but on weekends he may wear a custom-made red and blue wool, cotton, and linen uniform of an eighteenth-century Spanish artilleryman. On those days Brehm gets to perform one of the favorite parts of his job, shooting a six-pound cannon. And it is a performance that he and his team solemnly reenact in Spanish: the soldiers' ritually loading, lighting, and firing cannons from atop the fort.

The six-pound cannons use three pounds of gunpowder to fire a three-inch-diameter solid iron ball one and a half miles. The reenactors at the castillo shoot bread wrapped in tinfoil instead of a ball. It makes a tremendous sound, provides food for birds in the harbor, and makes you wonder how many Spanish soldiers went deaf or ended up with hearing problems. At maximum firepower the castillo had seventy-seven cannons, the largest of which shot twenty-four-pound balls, three miles at six hundred miles per hour.

"Hollywood gets it all wrong. Movies show cannons going *boom-boom-boom-boom-boom* all the time," said Brehm.[4] "In reality, they could shoot one shot every ten minutes. Plus, when these guys were holed up in the fort, they had to deal with how much gunpowder they had and not waste it." In addition to firing cannons, soldiers protected the fort by pulling up the drawbridge that spanned the moat and connected the ravelin to the main fort. The ravelin is the triangular outbuilding that served as guardhouse. It is the only part of the castillo that was never finished.

"One of the first questions most people ask after crossing the bridge," Brehm said, is 'Where's the water?' I always hate to disappoint them and tell them that once again Hollywood is misleading us. The moat was usually dry." When necessary, the castillo's moat could be filled as the tide rose and flowed through a pipe into the forty-foot-wide, eight-foot-deep space. After the United States acquired the castillo along with the rest of Florida in 1821, by treaty and not by battle—proving again that the pen is mightier than the sword—engineers filled in the eastern part of the moat and built up the sea wall to support cannons.[5]

From 1938 to 1995 the National Park Service flooded the moat daily.[6] The constant supply of water, however, weakened the west side of the fort, which was the last part of the construction and had been hastily erected. Wide vertical cracks began to appear in its two western bastions. The cracks are still visible, although mostly filled in with mortar.

Walking along the base of the castillo, Brehm continued to dispel myths. During the battle with Moore, the Spanish supposedly snuck out of the fort at night and dug out the balls that had stuck in the walls and shot them back at the English the next day. "The Spanish had no such luck; a ball hitting a stone wall, even one made of coquina, would be useless by being flattened on one side," he said.

People also think that the holes in the fort's walls are from those cannonballs or from a 1740 siege. Instead, pigeons probably made the bigger ones by enlarging weak spots in the coquina for nests. "They are the only birds that eat the fort. They look at it as one big cuttlefish bone and chip away at it and eat what they chip off," Brehm said. The birds eat the shells for gastroliths or to provide extra calcium for females during breeding season.[7]

Pointing to hundreds of small holes dotting the lower eight feet of several walls, Brehm said "Everyone thinks 'Oh, firing squad.' That's not what happened. They are kind of brokenhearted to find that out, too." During the Civil War, Union soldiers on guard duty would patrol from the upper level of the fort with live ammunition in their guns. Because explosive ordnance was stored in the fort, the soldiers couldn't take their guns back downstairs, in case one accidentally discharged. They could either shoot their guns, often into the fort's walls, or trade them for an unloaded gun with the next soldier on duty. Once a month a master armorer would pry out the balls and cast new ones. At least someone took advantage of the wall's absorbing capacity.

I wanted to chip off a piece of coquina or at least reach out and touch the bivalves, but park service rules forbid fondling. I had never seen any stone like coquina. Every block of the fort contains shells. Billions upon billions of shells, in particular a one-quarter-to-one-inch-long species called a coquina clam (*Donax variabilis*). Coquina means "little shell" in Spanish. Some blocks are pure *Donax* but most contain a mishmash of coquina, surf, ark, and Venus clams, along with cockles, bits of starfish, oyster, and quahog. Sort of a conchologist's dream.

Down in the moat were hanging gardens of ferns, grasses, and purple asters, which had taken root in the porous stone. The gardens covered the walls every thirty feet or so, wherever water drained scuppers from the courtyard roof. And the plants didn't just grow outside. In one of the courtyard rooms in the 1930s, the park service used to maintain a "fern room" almost completely covered in maidenhair fern. Now only a few ferns grew in this room.

The walls were plant rich because the coquina is shell rich. The heterogeneous mix of shells make a Swiss-cheese-like surface, where seeds can land and get established. Water accumulates in the cavities, further turning the coquina into a nursery. During a recent botanical survey, botanists found 153 plant species within the park, including 56, ranging from moss to elm, that had colonized the hanging gardens of the fort's walls. Cyanobacteria, nematodes, fungi, and diatoms have also established themselves on the coquina.[8] It is quite a cozy place.

Despite the beauty of the flowers, maintenance workers at the castillo constantly pull out the plants by hand. They don't want the roots to get established and do what cannonballs couldn't—destroy the fort. Clearing the walls of plants takes about six months. At least the plants grow fast enough that you can still see this wonderful link between geology and botany.

From the outside wall of the moat, Brehm crossed a twenty-foot-wide grassy area, called the covered way, to another coquina wall. This one rose another four feet; during the 1700s it would have been several feet higher. The wall provided a safe, or covert, zone in which soldiers could move around during battles. "No one could be seen. Guys would pop up out of the earth and shoot down the glacis. The wall created a human chain gun [or machine gun]," said Brehm.

The glacis (glah-*see*) was the open field that surrounded the fort. It sloped down from the covered way at such an angle that cannons from the fort could be tilted at their lowest angle and hit any enemy soldiers ascending the glacis toward the fort. Because the cannons shot in a straight line, the cannonballs would clear the men in the covered way. "With men popping out of the covered way and the cannons on the fort, it turned the whole glacis into a killing field," said Brehm. "The castillo really was the ultimate evolution of the military fort."

Pedro Menendez de Avilés established Florida's first permanent settlement, St. Augustine, on September 8, 1565. Before him had come Spaniards Juan Ponce de Leon, who sought youth and named Florida, and Hernando de Soto, who sought gold and introduced pigs to the New World, as well as Frenchman René de Laudonnière, who sought religious freedom and established a settlement, Fort Caroline, at the mouth of the St. John's River, about forty miles north of St. Augustine.[9] Menéndez simply wanted to rid Florida of the French.

Within two weeks he met with success by destroying Laudonnière's colony. The Spanish now ruled Florida, which was critical to the defense of Spain's colonies in the New World. By extending so far south, Florida controlled access to Mexico and put Cuba and Spain's other Caribbean colonies within easy reach by ship. Furthermore, the Florida Current, the beginning of the Gulf Stream, shoots north along the coast. The fast water was dangerous and sank many boats, but it also propelled Spanish ships and their gold along the fastest route back to Europe.

Menéndez's first act was to build a fort in a great lodge given to him by a native chief. Relations soured, however, and by April the natives had burned the fort to the ground.[10] Seeking safer territory, Menéndez moved his men to nearby Anastasia Island, where they erected a triangular wooden fort surrounded by a moat. This time the sea felled the structure. Another fort followed but mutinous soldiers torched it. In 1572 the Spanish moved back across the harbor and began building another fort.

Although they had probably encountered the widespread and abundant coquina on the island, the primary building materials there were rot-resistant cedar and cypress and tall, straight pines. Fort number four also succumbed to the sea. On the fifth attempt the Spanish tried to strengthen the fort by capping the high inner walls with mortar, made from lime (a binding agent) and sand. Six years later, though, it was "nothing more than a . . . storehouse for mice!"[11]

Although the next fort had the honor of being the first one named—San Juan de Pinos—it also had the dubious honor of shortest life. In early June 1586 global circumnavigator and pirate Sir Francis Drake arrived at St. Augustine. He burned the village and the fort to the ground.

With Drake's razing of town and fort, St. Augustinians embarked upon colonial America's first urban renewal.[12] They rebuilt the town and the fort—number seven—which received its long-standing name San Marcos. Again wood was the building material of choice, although a new governor had written the king in 1583 and suggested that coquina would make a good fort.

At long last recognizing that wood forts were unsafe and expensive to maintain, Spain approved a request for ten thousand ducats and twenty-four slaves to build fort number eight of coquina in 1595. Twenty slaves, one master stonecutter, one apprentice, and three masons were to arrive from Cuba.[13] On September 22, 1599, a storm surge tore away part of this new fort, which for reasons unknown had been built not of coquina but of "wood, sand and flour sacks."[14] Uncharacteristically, the rebuilt fort—number eight—survived, perhaps because slaves and subjugated natives regularly repaired and rebuilt the walls.

In addition to building with inadequate materials, Augustinians had to worry about a new potential threat when the English landed at Jamestown Island in May 1607. The new settlement's success guaranteed that England would spread out along the East Coast and eventually contend with Spain for dominion on the continent.

Fort number eight finally collapsed and St. Augustine fell on hard times during the mid-1600s. The garrison shrank to 150 soldiers, who had been reduced to foraging for roots. On May 28, 1668, a supply ship arrived just outside the harbor and, as was customary, signaled the harbor pilot to come out in his launch. The pilot found his countrymen onboard and signaled the good news to town with two gunshots. Everyone relaxed and celebrated that food would soon arrive.

When the pilot boarded the supply ship, he discovered that pirates under the command of Robert Searles had overpowered the crew. Searles and his men overwhelmed the unprepared town and took what little they could find in St. Augustine, though they didn't raze the town or the fort. Instead, they vowed to return with more men and ships and use St. Augustine as a base to capture trade boats sailing the Florida Current.[15]

Searles's action and proposed return scared the bejesus out of then

Florida governor Guerra, who wrote out a request to Spain for money and permission to build a stone fort. Guerra had one problem: Searles had taken all the sails necessary to equip a warship to sail to Cuba, the closest major Spanish port, so Guerra had to rely on the harbor launch to head out into the Florida Current and hope to meet a Spanish ship bound for Cuba. Captain Menéndez took the launch on July 8. It sank the next day and he walked back to town. A month later, after borrowing every scrap of sail that could be found, Menéndez sailed out in the lone warship. He reached Havana on September 9.

Havana provided some money and food, but Menéndez had to travel to Mexico and the capital of New Spain to get more money and permission to build a new fort. He arrived in Mexico City in early November 1669. Working with all diligent speed to defend Spain's most critical Florida settlement, the viceroy approved sending twelve thousand pesos and men from Cuba on December 16, 1670. In the intervening year the Spanish government had appointed a new Florida governor, sent three royal decrees saying that a fort should be built, heard rumors that St. Augustine had been destroyed (which delayed a shipment of soldiers and food), and debated whether the queen had expressly said to send money for the fort or merely thought it was a good idea. The queen's attorney in Mexico raised this last point.[16]

Outside events again affected St. Augustine. In April 1670 English ships landed at the Ashley River and established a small town, named in honor of their king, Charles II. Charles Town was only two days' sail from St. Augustine and within Florida's territorial border. In addition to pirates, the Spanish had to fear colonists.

Six months after the viceroy's approval, Florida's new governor, Manuel de Cendoya, sailed from Mexico and arrived in Havana on June 29, 1671. Havana would provide masons, stonecutters, and lime burners, plus military engineer Ignacio Daza, who would be in charge of building and designing the new fort. Cendoya finally reached St. Augustine on July 6, 1671, over three years after Searles had pillaged the town. Luckily the pirate didn't keep his word. Cendoya arrived with money and fifteen stoneworkers, but no Daza, who was not scheduled to come north until August 1672.

On July 12 Cendoya ordered work to begin on the fort.[17] He wanted 150 Indian laborers (peons) to assist the fifteen Cubans. Fifty men would quarry and transport coquina; fifty would make lime; and fifty would cut stone and dig the foundation trench. Blacksmiths made axes, pry bars, shovels, picks, and wedges. Carpenters built baskets, boxes, buckets, and square-ended dugout canoes. To make lime, the Spanish collected oyster shells and burned them in kilns, which drove off carbon dioxide and left behind a white powder.

At the quarry on Anastasia Island, the first task was to clear away the scrub oak, palmetto, and rattlesnakes and reach the coquina, which lay under a thin layer of soil. Peons used axes and picks to cut grooves deep into the soft stone. They pried out blocks with wedges and pry bars, taking advantage of layers of sand within the coquina that formed a natural splitting surface. A waterlogged block, two feet thick by four feet long, required six men to lift it. Workers transported the blocks by cart to a creek, where they loaded the stone on boards laid across the canoes. On the far shore north of the fort, more men unloaded the blocks, which needed to season for at least a year.

Workers also proceeded with lime production essential to making mortar. By March 1672 the kilns had generated sixty-three hundred bushels of burned-down oyster shells. They had most likely been collected from shell middens that dotted the coast. Native people, who had inhabited coastal Florida for thousands of years, relied heavily on oysters in their diet and had generated the shell mounds, which could stretch for several hundred yards in all directions and rise tens of feet.

Engineer Daza finally arrived from Cuba in midsummer. Governor Cendoya, Daza, and other dignitaries broke ground for the foundation trench at four P.M. on October 2, and on November 9 they laid the first block of coquina. Everything was ready for quick construction of the castillo made of clamshells.

Coquina around St. Augustine is like George Washington on the East Coast: everyone wants to claim a connection. You can live in a planned community at Coquina Crossing, bowl at Coquina Lanes, and pray at Coquina Community Church (based on a "more reliable foundation stone, that of Jesus Christ"). You can buy bags of coquina, too. The city visitor center offers a better deal than the national park. For $2.83 you get two marble-sized bits and a confusing fact sheet that calls coquina both a "natural limestone" and a "rocklike substance."

You can also visit three sites on Anastasia Island with signs stating that stone for the castillo came from that particular location. A coquina chimney and an old coquina well filled with trash stands near one sign. Nothing nearby looks like a quarry. Another sign stands near a swimming-pool-sized pond, a good spot to see great blue heron stalking small fish.[18] Again, no surface feature indicates that this might be a quarry. The third seems the most likely, with low slopes of coquina rising from a one-hundred-yard-long, fifteen-yard-wide shallow depression. It is also the best-marketed site, with prominent signs and two brochures describing its history.

Considering the size of the castillo—Joe Brehm said that a geologist estimated that over five billion pounds of stone had been used to build it—each of the signs is probably correct. What may be more surprising is that more sites don't make the same claim.

Site number three is on the northwest edge of Anastasia State Park, on the east shore of the island. A carved wood placard nailed to a slatted fence designates the area as "Old Spanish Coquina Quarry." Behind the sign runs a flat, shell-fragment-covered path, which curves through two hundred feet of scrub oak and palmetto into a clearing. If you arrive early in the morning, the sun may backlight mist rising from warmed grasses and low shrubs growing in the quarry bottom.

About sixty feet from the opening, a slope of tan to white shells and shell fragments weathers out of a ledge of coquina. Tucked into an overhang of ferns and shrubs, the top layers of fresh coquina are peachy to orangish tan, a result of oxidation of iron in the stone. Most of the wall is the more typical tan, which subsequently bleaches out in the sun to the gray of the castillo.

The drab colors of coquina, the stone, fail to convey the beauty of coquina, the clam. As the specific name, *variablis*, implies, no two look alike. Never uniform and always complex, the colors can be banded, like growth rings, fingered, like rays of sunlight, or both. Set against a yellow, ivory, or white background, rays and bands can be yellow to russet, blue to purple, pink, gray, or any mixture of these shades. Such variation has led to common names of butterfly shell and periwinkle.

The *Donax*'s kaleidoscope of pattern and color helps make the clams less visible to predators, such as oystercatchers, sanderlings, and ghost crabs. One way to locate coquina clams is to look for groups of birds probing the sand to find the buried bivalves. Less mobile predators include moon snails, which nab *Donax* with a lightning dart of their foot, and lettered olive snails, which latch onto a coquina and pull it under the sand for a subsurface snack. Fish also eat coquinas or at least have the "nasty habit of biting off the siphons," says marine biologist Olaf Ellers.[19] Coquinas have two siphons: one for sucking in food, the other for expelling waste.

People also eat coquina clams. The earliest evidence of people inhabiting what is now Florida comes from coquina shell middens fifty-seven hundred years old. Ancient beach dwellers appear to have eaten coquina clams seasonally for thousands of years. Archaeologists don't know exactly how prehistoric people ate coquinas, but modern molluskivores prefer a broth, which writer Marjorie Kinnan Rawlings describes as delicate and delicious.[20] She recommends serving it piping hot with "two tablespoons thin cream and a small lump of butter."

At the quarry, away from the prying eyes of law-abiding, law-enforcing national park rangers, you can run your hand over the coquina. Shells will spill out and cascade down the slope. In some layers, roots poke out, having penetrated over a foot deep. They look as if they would slice through the rock if you pulled on them. How could anyone think this clammy material would be good for building, particularly for a fort to protect a town from rampaging pirates and overzealous Englishmen?

Known officially as the Anastasia Formation, the 110,000-year-old layers of coquina outcrop as a wisp along the east coast of Florida from Anastasia Island south to Boca Raton. Stretching for 225 miles, the rock

Close-up of coquina showing shells.

formation is never wider than 10 miles and seldom thicker than six feet, although some beds are up to fifty feet thick.[21] Made primarily of shells, shell fragments, and quartz sand, the Anastasia can be seen at Marineland, Flagler Beach, Blowing Rocks Preserve, and House of Refuge (in Martin County).

Coquina is geology in its most elemental form: not quite rock and not just a pile of shells, more a stone during its gestation. It is sort of the geologic opposite of Morton Gneiss. Every species found in the Anastasia can still be found living on beaches in Florida. No life existed on Earth when the gneiss formed. In the gneiss you can see how billions of years of geologic processes have melted, mixed, and metamorphosed the rock, whereas the coquina doesn't appear that different from the day it first formed. Look at gneiss and you can tell it is very hard rock. Touch coquina and you can feel that it is very soft rock. Part of the appeal of coquina, at least to a modern visitor, is its elemental simplicity. You can easily understand how a stone forms.

Imagine a beach, any beach where clams and cockles and scallops and crabs live. They die. In places, the currents sort the remains into homogenous piles of small clamshells. In other locales water mixes the shells into heterogeneous heaps. Time passes. Rainwater washes down into the piles and dissolves away the calcium carbonate that makes up the shells. Gravity pulls the water down into the pile. The water becomes saturated and begins to deposit minute crystals of calcium carbonate on any surface the water touches. The crystals cement the shells together and coquina is born.

The new coquina changes little for hundreds of thousands of years, but if we peer into the future, we would eventually see new materials, such as sand, silt, and even more shells, beginning to accumulate on the coquina. The weight would compress the layers, which would become denser. More cement would accumulate and after millions of years, the coquina would now look like a rock, solid and hard. You would still recognize the fossils but no original calcium carbonate would remain. The coquina would now be a true limestone.

In the Anastasia, cementation, or lithification, was aided by the *Donax* shells. Like most bivalves, *Donax* shells are made from calcium carbonate, but not the usual form of calcium carbonate, called calcite. Instead, *Donax* are made from aragonite, a less stable form of calcium carbonate. If *Donax* had been made from calcite, rainwater would not have been able to dissolve the calcium carbonate as quickly, and the shells may not have been cemented together into coquina. The Spanish might not have retained Florida and life as we know it may have been completely altered. Fortunately the clam was there.

*Donax* clams also affect coquina after the stone has been quarried. In contrast to how brownstone seasons—by internal water transporting minerals to the stone's exterior where calcite and silica reprecipitate on the surface—coquina hardens through external water, either rain or runoff, dissolving aragonite and precipitating cement near the surface. In both cases, seasoning creates a harder shell, sort of like chocolate coating on an ice cream bar.[22]

Lithification occurred and still occurs very rapidly in the Anastasia Formation. South around Cape Canaveral, perfectly preserved ghost crabs, which died in their burrows, have been frozen into the coquina. They look like some twisted version of Shake 'N Bake with grains of sand, shell bits, and calcite coating the crab carapace.[23] Closer to St. Augustine, lithification has locked Coke bottles in place.

Deposition of the Anastasia occurred in a high-energy, shallow marine environment. Specifically, the coquina accumulated in the swash zone, the area along a beach intermittently covered and uncovered by waves. If you think about any swash zone you have seen, you will realize that it is a harsh environment, with waves constantly pounding and abrading the beach, which helps explain the abundance of shell fragments. And because of currents and storms depositing sediments, layers of nearly pure sand periodically interrupt the layers of shells, which created the zones of weakness that quarry workers exploited in order to split the coquina into blocks.

The swash zone is an ideal habitat for coquina clams, says Olaf Ellers, who wrote his Ph.D. dissertation on the mollusk's migratory movements. Winter finds *Donax* offshore in deeper and safer water, away from predators. During warmer months, they follow daily tides, moving inland with a rising sea and seaward as the water retreats. Their goal is to situate themselves in the swash zone where moving water suspends food particles. Once they reach their ideal location they use their shovel-like foot to dig themselves into the sand. Bad weather doesn't stop coquinas from their travels, as Ellers found out when he snuck into a hurricane zone while working on his dissertation and watched coquina battle the big waves. I don't recommend following Ellers's lead.

To facilitate movement, coquinas pump their foot up and down like a pogo stick to leap into the surf. Ellers coined the term "swash-riding" to describe this molluscan leaping and surfing. Like their human counterparts, coquinas are picky about their waves. "They choose only the 20 percent of waves that would carry them the furthest distance shoreward," he says.

Ellers further discovered that they detect which waves to surf by feeling sound vibrations generated by the crashing waves. "It's analogous to the vibrations you feel when a large truck passes," he says. "Larger waves produce more vibrations." The coquina are so in tune with these sounds that Ellers learned to play a trick with them. In his lab, he would fill a bucket with bivalves and impress his friends by playing a sound and getting all of the clams to jump.[24] Conchologists must be an easy group to impress.

When the Anastasia formed 110,000 years ago, Earth was in what is known as an interglacial period, one of a number that have occurred during the past 2 million years. We are in an interglacial at present, meaning that the planet is between ice ages, the last of which ended about thirteen thousand years ago. The big difference is that during Anastasia times, the climate was warmer than it is at present, which melted more of the planet's glaciers and made sea level as much as twenty-five feet higher than it is today.

As temperatures cooled during the ice age following the Anastasia interglacial, sea level began to fall, eventually dropping three hundred feet below modern sea level during the last ice age, as the glaciers tied up water that normally would have been in the oceans. Falling sea levels stranded the recently deposited shells far inland. Freshwater soon mixed into the deposit and began to cement the shells. The Anastasia Formation was born.

In 1596 it became the first building stone used by European colonists in what would become the United States. The Spanish used coquina to erect a powder magazine, a building that housed gunpowder, in fort number eight.[25] The oldest extant coquina structure is a well, built around 1614. Nothing else appears to have been built from coquina until the castillo.

What makes coquina more remarkable is where it outcrops. If Pedro Menéndez de Avilés had decided to establish his base of operations north of St. Augustine in 1565, his late-seventeenth-century followers would have been in trouble. No good port with access to freshwater and good building stone occurs along the coasts of Georgia, South Carolina, or

North Carolina. Serendipitously, Menéndez had chosen the one harbor with good stone. Without those *Donax* clams, who knows what would have happened to the Spanish?

"The castillo was a fantastic deterrent. Anyone sailing by sees this little harbor, little town, and great big hulking white and red fortification, immediately knows this is a Spanish fort with lots of guns," said Joe Brehm. "It says, 'Don't mess with us or we will hurt you.' "

The design was not unique. After Ignacio Daza surveyed the site, he pulled out his military engineering books, which contained a catalog of designs, and "literally found a fortification that would fit his needs but scaled it down," said Brehm. "The castillo is essentially a one-fifth scale model of what was known as a 'frontier fortification.' If you looked at a map of Paris and the surrounding area in the seventeenth century, you would find at least seven forts that look just like this one."

Military engineers in the sixteenth and seventeenth centuries based their designs on siege warfare, where the defenders hunkered down in their fort and the attackers attempted to get inside. Success depended on cutting off supplies to the hunkerers, destroying the fort's weapons, particularly cannons, and breaching or scaling the walls. Sieges could take months and result in demoralizing losses for the attackers. "Frederick the Great normally had casualties of fifty-five to sixty-five percent in a siege. There just weren't enough people in Florida to make a good siege," said Brehm.

With a design in place, Daza began to supervise work on the castillo.[26] Although illness weakened the Indian peons and forced Governor Cendoya and his soldiers to man the shovels, the north, south, and east walls began to rise. By early 1673, the east wall and two bastions stood eleven feet high. Then Cendoya and Daza died within days of each other, the viceroy in Mexico refused to send money until ordered by Spain, and a storm breached the stone walls and wrecked the old wood fort. Construction also slowed because a ship carrying provisions foundered and workers had to go in search of food.

Despite the challenges, the fort grew. In May 1675, when a new governor, Pablo de Hita Salazar arrived—with food, too—the south wall and southeast bastion were up to a height of twelve feet, the north wall and northeast bastion to twenty feet, and the east wall to fifteen feet. The walls were thirteen feet thick at the base, tapering to nine feet at the top. Recognizing that he needed a defensible fort, Hita decided to erect a temporary, twelve-foot-high wall of dirt, wood, and stone veneer on the west side between the half completed northwest and southwest bastions. He also finished the three other curtain walls and seaward bastions, and mounted cannons on the triangular structures. St. Augustine finally had its first real fort.

In a report back to Spain, the governor wrote of the castillo that "in the form of its plan this one is not surpassed by any of those of greater character." He also observed how little food and money he had to provide his workers. "If it had to be built in another place than St. Augustine it would cost a double amount . . ."[27]

Governor Hita could make this observation because of the international—that is, mostly subjugated—nature of the workers. Guale, Timucua, and Apalache natives joined Spanish peons, African slaves, Caribbean convicts, and English prisoners, along with Cuban and Mexican mestizos and St. Augustinian creoles. Wages ranged from zero pay and limited rations for slaves and convicts through one real (twenty cents) per day and rations for Indians, to twenty reales for the master mason. There was room for advancement: English prisoner John Collins worked his way up to master of the kilns and eventually to quarrymaster.

Hita's initial enthusiasm notwithstanding, fort construction halted on December 31, 1677. The viceroy's parsimony left no money to pay workers; during the last months of 1677, work continued only because of gifts from town residents. Work did not begin again until August 29, 1679. Ever so slowly the walls of the castillo rose, with periodic fixes on older sections that had been poorly or incorrectly built.

The castillo's first test came in spring 1683 when 230 pirates landed at Ponce de Leon inlet, sixty miles south of St. Augustine, and began to march north. They easily overpowered a watchtower on the south end of Anastasia Island, but when they reached the north end of the island the pirates realized the stone fort meant that taking the town would no longer be as easy as it had been. As Moore would do nineteen years later, they abandoned their nefarious plans and decided to retreat.

Flushed, or perhaps shocked, with St. Augustine's first victory, the castillo crews pushed harder to finish and by 1685 all interior work was done. This included the interior courtyard, now surrounded by more than twenty rooms with wood support beams and a tabby (oyster shells and mortar) slab roof.[28] All that remained was finishing the ravelin, the moat, the covered way, and the glacis. Without these critical defensive outworks, the fort was still vulnerable to a siege attack. Furthermore, the wood-beamed rooms could not support the heavy cannons, which could be used only on the thick, coquina-walled bastions.

International politics again affected St. Augustine. Taking time out from working on the fort, the Spanish tried to crush the recently established colony of Charles Town, but a storm stopped them. They also had to battle English advances in western Florida and Georgia. And then Spain declared war on France, which gave French corsairs more reason to attack Spanish supply ships bringing food to St. Augustine. Food became so scarce that officials decided to plant corn on the glacis surrounding the castillo.

Finally in August 1695, the castillo and its outer defensive works, except the ravelin, were completed. Plastered a brilliant white with red watchtower and garitas, little towers at the corner of each bastion, the castillo had cost 138,750 pesos, about double what Governor Cendoya had estimated in 1672. Seven years later, the new fort proved its worth during the defeat of Moore.

Completion of the castillo and Moore's razing of the town created the market for coquina during St. Augustine's second phase of urban renewal. Residents, at least the wealthier ones, used coquina for houses, wells, and garden walls. The Catholic Church also chose coquina, as did the Spanish government for their official buildings.

Despite the Spanish word, no Spaniards used the term coquina to describe their building stone. They preferred *piedra* (stone), *canto* (quarry stone), *canteria* (hewn stone), or *mamposteria* (stone masonry). When the English took control of St. Augustine in 1763, they called the coquina "stone" or "shellstone." The earliest known reference to coquina is from 1819. Not until the late 1830s did the term start to catch on.

Some early visitors to St. Augustine didn't dignify coquina with the term "stone." An English traveler in 1817 wrote, "This marine substance is superior to stone, not being liable to splinter from the effects of bombardment."[29] In 1831 John James Audubon wrote to his wife "an old Spanish Castle . . . is built of . . . a concrete of shells which hardens by exposure to the Air and is curious to the Geologist."[30]

The modern town of St. Augustine has recognized the importance of coquina to its past and has worked extensively on preservation, restoration, and upkeep of its historic structures. It has also attempted to keep the spirit of the Spanish architecture, at least downtown, with narrow streets overhung with second-story balconies. Unfortunately, many merchants seem to miss the point and use the historic aura as a means to sell cheesy tchotchkes.

Anastasia Island remained the primary source for quarries. They spread in a narrow band along the center of the island. None were large and none were deep. The Spanish government owned the quarries and controlled all access. Except for the years 1763 to 1783, Spain retained ownership of the quarries until 1821, when the U.S. government acquired Florida. It also acquired the quarries and surrounding land, which is how the land now run as Anastasia State Park stayed protected and undeveloped.

Thirty-eight years after Moore's failure, the British tried another attack on St. Augustine and the castillo. On June 13, 1740, fourteen hundred men under the command of Georgia governor James Oglethorpe began to bombard the fort. They were better prepared with more men, bigger guns, and mortars, hollow shells lobbed high to explode over or on the grounds of the fort (Francis Scott Key's "bombs bursting in air"). The coquina withstood the onslaught, leading one British soldier to express "[it] will not splinter but will give way to cannon ball as though you would stick a knife into cheese."[31] Oglethorpe and his men retreated on July 16.

With peace at hand, the Spanish began to beef up their defenses. They built a new coquina structure, Fort Matanzas, at the south inlet to the Matanzas River. They rebuilt the coquina wall of the covered way; remodeled and finished the four defensive lines that formed successive walls around St. Augustine; and expanded and converted the rooms surrounding the castillo's courtyard by replacing the wood beams with coquina and tabby vaulted ceilings, which shrank the courtyard by fifty feet and raised the castillo's walls five feet to thirty feet above the moat bottom. For the first time, the Spanish could place cannons anywhere along the fort's walls, not just atop the corner bastions. By 1762 all except the ravelin was done, including a new coat of plaster.

Ironically, a year later the French and Indian War, also known as the Seven Years' War in Europe, ended with victory for England, which acquired Canada from France and Florida from Spain. St. Augustine and the castillo, now called Fort St. Mark, transferred to British control on July 21, 1763. The town became a British haven: the only one of the colonies to support the British during the Revolutionary War.[32] Twenty years later Florida reverted back to Spain as part of the treaties signed in Paris marking the end of the American Revolution.

Daily life in St. Augustine continued to be a struggle with limited food (commonly eaten items were mullet, catfish, and gopher tortoise) and delayed subsidies from Spain, although new governors would arrive and bring energy for construction and increased trade and the town would prosper.[33] One big change from the seventeenth century, however, was that Augustinians now knew that their little town was permanent; they could always depend on the castillo as their fortress, not only as a deterrent but also as a refuge. St. Augustine also started to get what might be called its first tourists.

"Augustin itself is widely known to be a healthy place, so that weaklings and consumptives from the northern provinces resort hither, and always to their advantage," wrote German botanist and explorer Johann David Schoepf.[34] Schoepf made his observation during travels in Florida in the early 1780s. Time, however, did not treat the town well. When John James Audubon visited in 1831, ten years after the United States acquired Florida, he wrote, "St. Augustine is the poorest hole in the Creation."[35] The castillo, or Fort Marion, as the Americans named it, must have impressed him because he used it as the backdrop for his painting of the Greenshank, even though he shot the bird at the southern tip of Florida.[36]

Although no shots flew, the South (1861–62) and the North (1862–65) each controlled the fort during the Civil War. After fighting ceased, Fort Marion served as a prison for Great Plains and Apache Indians. The grounds later became Florida's first golf course, when Henry Flagler, cofounder of Standard Oil, built his stunning Hotel Ponce de Leon in St. Augustine. The former hotel, now Flagler College, was the first large-scale building constructed entirely of poured concrete; the walls look like coquina because the builders used crushed coquina as an aggregate.

Flagler also built a railroad to town and began to promote St. Augustine as a tourist destination. The quiet hamlet became the "Newport of Florida," as Harriet Beecher Stowe described it.[37] Visitors could buy coquina carvings and painted coquina from the "Coquina Man." The St. Augustine Historical Society offered tours of the castillo. Because the guides depended on tips, the stories became more fanciful with a torture rack, quicksand pit, and secret dungeon taking over where truth ended.

The arrival of Flagler's railroad had another consequence. Now that the railroad could bring any stone builders wanted, coquina faded as a building material, not to reappear again until Franklin Roosevelt's New Deal projects, which emphasized the use of local materials. In St. Augustine, the

New Deal led to construction of a civic center and a large hotel. Daytona Beach also incorporated coquina in two magnificent structures: a bandshell, now hidden by beachside condos and hotels, and the Tarragona Arch, which on the day I visited was decorated with Christmas lights outlining a motorcycle-riding Santa.

Only one quarry now supplies coquina. Located about four miles west of St. Augustine, it is deeper than the surface deposits of Anastasia Island. Owner Gary Wilson says that he supplies stone for statues such as dolphins and turtles, and veneer for new developments. Clients have included two people one might not suspect of sharing a common interest: actor Burt Reynolds ordered a carved Brahma bull and golfer Vijay Singh wanted coquina veneer for his home and guesthouse. Coquina truly does bring the world together.[38]

Wilson also supplies stone to the castillo. Over the past few years, the park service has been engaged in a multifaceted restoration process. Workers have experimented with chemical methods to prevent plant growth on the coquina. They have rebuilt sections of the covered way stone walls and resurfaced the roof on top of the courtyard rooms to prevent water damage, although water seepage continues to produce stalactites in some rooms. Park employees are also using Wilson's coquina to stabilize horizontal surfaces in the fort.

Although most of these restoration projects focus on preventing water damage, the park service has no plans to do the one thing that would best protect the fort and make it historically more accurate. They will not replaster the building and restore the historic white and red colors. They claim that people would complain because they are used to the gray color of the castillo. There is probably another reason. They just don't want to cover up the clams.

# 6

# America's Building Stone—Indiana Limestone

*"We have under our feet the best building material God ever put on Earth. Because of that fact, this industry is as nearly eternal as you can get. One hundred years from now, people will still be hauling limestone out of this little patch of ground. They may be shipping it on spaceships and light rays, but one way or another they'll be hauling it out of the ground and stacking it into the air."*
—Bill McDonald, as quoted in *In Limestone Country*,
Scott Russell Sanders

*I cut the stone for this building . . . I was proud of my work. When they were finished the darndest thing happened. It was like the buildings were too good for us. Nobody told us that . . . it just felt uncomfortable.*
—Mr. Stoller, in the film *Breaking Away*

IN SEPTEMBER 2007 I went in search of limestone. My destination was Bloomington, Indiana, the heart of a region where limestone has been quarried for almost two hundred years. The stone was everywhere. Green slate topped a few of the walls but basically every rock was gray or buff limestone. The highest rose sixty or seventy feet, smooth and cool to the touch. Black lichens covered the north sides of these clifflike faces and were especially prominent in areas where water had trickled down. Other walls made of broken slabs of limestone reached only as high as my waist. On one an untidy line of flagstones stood perpendicular to the horizontal slabs, looking like an upside-down photo of James Hutton's famous unconformity at Siccar Point.

At the base of the slabs, moss and ferns pushed out of crevices and added a beautiful verdant contrast to the buff blocks. Towering above this low wall were oaks and maples with a dozen red admiral and question mark butterflies sipping sap from the trunk of one oak.

On another block of limestone someone had carved an intricate pattern of vines and leaves. Whoever made the design used a narrow chisel, its cutting grooves still visible in the soft stone. Nearby the date *1890* had been chiseled eight inches high. The numbers were remarkably crisp and sharp considering the area's classic rock-unfriendly climate of hot, humid summers and freezing winters, which typically ravage limestone.

Walking away from one high wall of rock, I followed a leaf-covered path through a wooded area of beeches and maples to a grassy field and a low arch of smooth limestone. When I got closer, I noticed fossils crowded together in several blocks of rough stone on either side of the opening. Afternoon light hit the fossils obliquely and they stood out from the softer substrate like minute tombstones. This graveyard of invertebrates entombed brachiopods, one- to two-inch-wide, clamlike shells; crinoids, consisting of a cuplike calyx and thin ridged discs, some stacked five or ten high; and bryozoans, which resemble broken bits of Rice Chex cereal.

Close-up of Salem Limestone, Maxwell Hall (1894), Indiana University campus, Bloomington, Indiana.

Brian Keith, a geologist with the Indiana Geological Survey, calls these three animals the "holy trinity" of the Mississippian Period.[1] Brachiopods, bryozoans, and crinoids were some of the most abundant invertebrates living from 354 to 320 million years ago, when a shallow, warm sea covered much of what we now call the Midwest and deposited the limestone. With a 10× hand lens, I could see that the wall of fossils was completely made of shells, most of which had broken into pieces in the fast moving tides that daily swept the ancient sea.

Although I would have liked to smash off a sample of this fossiliferous rock and take it home with me, I knew I couldn't because I was not in a wilderness or a semiwild spot. I was on the Indiana University (IU) campus in Bloomington and I was sure that the Hoosier faithful would look askance at someone, especially someone born in Kentucky, pecking away at one of the oldest buildings on campus.

All but a few of the buildings on the IU campus, as well as many of the offices, banks, and government buildings in Bloomington, are made from the stone menagerie known variously as Bedford, Indiana, oolitic, or Salem limestone. The stone comes from a thirty-mile-long-by-five-mile-wide area called the Indiana building stone district. The Belt, as those in the trade call it, stretches northwest from Bedford to Stinesville, about ten miles northwest of Bloomington. Workers first used the stone in 1819, in the foundation and windowsills of the Monroe County Courthouse. The men hauled the blocks eight miles, ironically along an area later dotted with quarries and to a site resting upon extensive beds of limestone. The first quarry opened eight years later and builders have used Indiana limestone more or less continuously ever since.

When you walk through Bloomington and Indiana University you are in the center of the Salem Limestone universe—builders in town tout limestone countertops instead of granite—but no matter where you are reading this book you do not have to travel far to find what Brian Keith describes as the "premier building stone in the country." Salem Limestone may face your local government offices, make up the windowsills on a university campus, enclose the entryway to a bank, tower above as fluted columns in a courthouse, or accent the dark granite on a high-rise office building. As far as I have been able to determine, the Bedford rock is the only building stone used in all fifty states.

The first building stone one of your relatives encountered in the United States was probably Salem Limestone, too. The off-white, fossil-rich stone trims walls and doorways at the immigration station at Ellis Island, built in 1900. Open until 1954, Ellis Island welcomed 12 million immigrants; over 40 percent of all Americans can trace their ancestry to those who walked through the island's Salem-framed doorways. Perhaps a few geologically inclined emigrants backed up the line as they gazed thoughtfully at the fossils.

You are probably even carrying with you a picture of a building that incorporates Salem. Reach into your wallet and pull out a bill. If it is a five, a twenty, or a fifty, turn it to the back and look at the building. The White House and U.S. Capitol, although not built originally of Salem rock, have used the limestone extensively for repair work, and Abraham Lincoln sits on his marble chair surrounded by Salem Limestone.[2] And if you don't carry bills, then you have probably mailed a letter from one of the more than 750 post offices built with Salem blocks or perhaps paid a fine, obtained a marriage certificate, or watched a legislative session at one of the more than two hundred Salem-sheathed courthouses or twenty-seven state capitols.

"So many monuments and landmarks are made from the Indiana limestone that it is a holder of American memory," says limestone sculptor Amy Brier.[3] She is right. Other stones are older, more beautiful, and have more noble pedigrees, but no other building stone forms as much a part of the collective cultural fabric of the United States as the Salem. No other stone has contributed more to giving our cities and towns a sense of elegance and pride. No other stone deserves to be called America's building stone.

Three hundred and thirty million years ago, during the Mississippian Period, you could have sailed a boat across most of the middle part of North America. You would have floated over future Illinois, Kansas, Iowa, Colorado, Arizona, and Indiana, though you would not have moved fast because you were in the windless zone of the globe we now call the Doldrums. In addition, you would have needed plenty of sunblock, as your boat would periodically cross the equator, which, because North America tilted almost 90 degrees to the northeast, ran from about modern-day San Diego through Duluth, Minnesota. In many areas, particularly around Indiana, the water was less than twenty feet deep.

Geologists know all this because the equatorial sea deposited sediments preserved in rocks across the country. At the Grand Canyon, a several-hundred-foot-thick layer of rock known as the Redwall Limestone formed in this shallow sea and shares the Salem's brachiopods, bryozoans, and crinoids. In eastern Colorado, geologists refer to their Mississippian Sea limestone as the Spergen, but in Kansas and Illinois the name reverts to the Salem Limestone, named from early quarries near Salem, Indiana.

As you sailed along, you could have traveled north and west for many days without seeing any land. Most of North America spread to the east and northeast, as a lowland now called Wisconsin and Canada. Far to the southeast, the eroding Acadian mountains rose out of the water. Staying in the south but moving west, you would have sailed off the platform of shallow water into a deep basin. The first landmass you would have seen was Gondwana, inching toward a collision with North America. Another range of mountains was also pushing up to the west, running at about a thirty-degree angle northeast from the equator. These mountains now stand in Nevada and are known as the Antlers.

Sailing along you would have noticed an additional facet of the water. Like the Bahamas, where modern limestone forms in a similar environment, the water would have been extremely clear because little or no sediment washed into the Salem sea. Any material that washed off of the mountains ended up in deep marine basins adjacent to the land.

The clear, warm, shallow water resulted in two characteristics of the Salem. One, the building stone section is nearly pure calcite, or calcium carbonate. All of the organisms that lived in the water had calcium carbonate skeletons or shells, and any sediment that accumulated consisted of calcium carbonate. Two, many of the sediments were round, like fish eggs. Known as oolitic grains, they formed when wave action rolled a particle and surrounded it in concentric layers of calcium carbonate, like what happens when you roll snow to build a snowman. Over the millions of years the sea covered the continent, it generated enough calcium carbonate to build up a ninety-foot-thick layer of limestone in Indiana.

If you had chosen to land on the shoreline abutting Indiana, the world would have seemed depauperate. No birds or mammals would have existed and the first dinosaurs were still almost 100 million years away. A few four-legged amphibians had traipsed out of the water, but scorpions, mites, spiders, and a host of insects would have dominated the terrestrial fauna. Ferns and low trees, including early conifers, would have formed extensive forests. Another 200 million years would have to pass before you could see the lovely deciduous trees that now flourish in Indiana.

Getting back in your boat on the shoreline, you would have sailed in a quiet lagoon, with a bottom covered in fine carbonate mud. Numerous invertebrates seeking food and building homes would have churned up the sediment and left faint traces of test-tube- and inverted-Y-shaped burrows, now preserved in the Indiana rock. If you climbed overboard and dropped into the oozy bottom, you could have collected a handful of mud, and if you had a microscope, you would have seen that although dead bodies made up most of the ooze, other life forms lived in the lagoonal graveyard. The microscope would have revealed a world populated by protozoans inhabiting one-twentieth-of-an-inch-wide shells, each made of a half dozen chambers coiled like a poorly made cinnamon roll.

Known as foraminiferas, they lived for a few months, died in the lagoon, and settled amid the billions of shells of their cohorts.[4] Forams are an abundant fossil in some parts of the Salem, but because of their wee size they are rarely visible in the stone. When you see a Salem wall you are looking at a cemetery of epic proportions.

Floating out of the lagoon, you would have passed into a shoal complex that spread along the coast for tens of miles. These underwater sandbars and associated channels, called intershoals, would become the main rock units quarried for building material in the Belt. They were high-energy environments, shaped by daily tides, longshore currents, and periodic storms.

"We are now able to see primary sedimentary structures in the quarry walls. There are no other exposures in the world where you can see this level of sedimentary detail," said Todd Thompson, another geologist at the Indiana Geological Survey.[5] A change in cutting technology in the past dozen years has allowed Thompson and his colleague Brian Keith to trace the ancient ripples and troughs of the shoals. They follow the dune migrations in the protean sea by holding chalking parties where students mark features of individual shoals and intershoals at the quarries. "We have to move quickly because the quarrymen are pulling the stone out so fast that the evidence disappears almost immediately," said Thompson.

By tracing out the bedding surfaces, the geologists can see the three-dimensional architecture of the ancient sandbars better than they could in any modern environment. At one quarry, Thompson found bedding with distinct thickening and thinning in groups of fourteen, which he interpreted as corresponding to fourteen days of deposition during the Mississippian Period.

One of Thompson and Keith's goals is to understand the geometry of the shoals so that they can help quarry owners quarry the best stone with the least amount of waste. Despite their best intentions, however, "There is no geology in the stone business," said Keith. Because habits and tradition guide the quarrying process, the quarrymen cut the walls in a rectilinear grid regardless of the geologists' advice that the shoal bedding doesn't follow precise lines.

Look at most any building with Salem rock and you will see Thompson and Keith's complex pattern of bedding, with some flat, some angled, and some concave upward beds. I say *most* because the most expensive stone has fewer recognizable geologic features, which may make for "better" building material but more boring rocks, in the eyes and hearts of geologists.

Quarrymen also eschew large fossils, another favorite feature of geologists. Fossil-rich blocks often got tossed to the side or dumped into old quarry holes because fossils made the stone less homogenous looking, although such blocks make just as good building material as blocks made of ground-up fossil parts.

Not all fossil-rich limestone got pitched aside; if it did builders would have little to work with. Many years ago, or so the story goes, two Chicago women traveled to Bloomington to look at limestone. They came across several discarded blocks loaded with fossils. The quarry owner told the women the stone was very rare and expensive. They bought the blocks. Ironically, one of the more highbrow social clubs in Seattle incorporates some of the better fossil-rich blocks of Salem Limestone that I have seen. Not wanting to cause a social brouhaha in my hometown, I haven't pointed out to the fine folk of the club that they have what builders consider "inferior" stone.

Few of the fossil animals found in the Salem building stone inhabited the high-energy shoal environment. Instead, they lived farther out to sea, in slightly deeper and quieter water. If you had pulled yourself away from the microscope you could have swum down to the sea bottom and found Brian Keith's trio of Mississippian beasties forming vast communities seaward of the Salem shoals.

Crinoids, also known as sea lilies, would have been the most obvious. They are one of the classic fossils found in limestones around the world. Widespread and abundant for hundreds of millions of years, only one group of crinoids has survived to the present and they now inhabit deep water. Crinoids anchored themselves to the substrate by a rootlike structure called a holdfast, out of which extended a stalk of stacked disks. The stalk supported the body, which consisted of a cuplike calyx and arms, usually 10 but ranging up to 250. The arms, which helped catch suspended food particles, were flexible and often broke. When they did, two new ones emerged. A fully erect Salem crinoid, what one crinoid specialist called a "a feathery starfish on a stick," could be twelve inches tall.[6]

Cruising along the sea bottom, you also would have seen housing colonies of interconnected rooms that looked like ice-cream cones anchored to the substrate, although some would have looked like an ice-cream cone twisted into a corkscrew. Inside each pinhead-sized room would have lived a tubular bryozoan. Water flowing across the netlike structure provided food for the hundreds or thousands of tentacled bryozoans that formed the colony. The delicate fronds might be compared to brownstone row houses—squat and extensive—relative to the sleek crinoids that towered above the three-inch-tall bryozoans.

You wouldn't have had any problem locating the final member of the trinity. Brachiopods look like clams, but like bryozoans, to which they are closely related, they wouldn't have moved. Often called lamp shells, for their resemblance to bowl-like oil lamps, brachiopods were one of the most abundant invertebrates of the marine world and could form dense colonies. At least twenty species of brachiopod dwelled in the Salem waters.

The Salem sea, however, was not an inert world populated only with immobile invertebrates. Numerous snails slithered across the seafloor, periodically stopping for a bite off a brachiopod or bryozoan. And you could have been a meal for the sharks that plied the waters. The largest ones had teeth over two inches long and bodies the length of an SUV.

When the invertebrates died, and they died by the billions, currents would have transported them out of the deeper water up toward the shoals and channels. Wave action then blenderized the bodies as shells and exoskeletons crashed against each other and broke into hundreds of pieces. The battered beasts piled up in an unsorted stew of body parts, comparable to the coquina used at Castillo de San Marcos, except the Salem sediments were finer grained.

Over time the unconsolidated Salem shells and skeletons compressed under the weight of more bodies and became denser and harder. What had started as a coquina had been converted over millions of years to a limestone. You can still recognize the shells and skeletons, but they look like fossils and not like recently dead animals, as they do in the walls of the castillo.

Salem Limestone shares one other characteristic with the Florida coquina. Cutters, carvers, and sculptors can work the Salem in any direction. The rock is so pure and homogenous that its bedding has little effect on how one shapes the stone. No matter how quarrymen slice or carve the rock, it has the same strength and durability. This feature, more than any other physical asset, helped make the Salem popular with everyone from Odd Fellows to opera fans, or at least to the people who designed buildings for them.

Historians recognize one Richard Gilbert as the first to quarry Salem Limestone. Starting in 1827, his small ledge along a creek at the northern end of the Belt provided stone for chimneys, monument bases, and bridge piers. Toward the southern end of the Belt, Dr. Winthrop Foote established the quarrying industry in Bedford in 1832, when he hired a man named Toburn, a stonemason from Louisville, Kentucky. Toburn's best-known work is the Foote family tomb, cut directly into a ledge of limestone. The tomb, reached by taking a side road east of town and walking a hundred feet down a beer-can-strewed path in the woods, no longer has crisp edges but has otherwise weathered well its many decades of rain, snow, ice, heat, and parties.

Like most other building stone, Salem Limestone remained a local product during its early decades of use. Not until 1853, when a rail line extended seventy miles south from Louisville, did much stone leave the Belt. By 1870 good railroad connections had been established north to Chicago and east to markets such as Boston and New York, and by the end of the century, train tracks snaked across the Belt, pushing in or near to dozens of quarries.

Early Salem quarries were primitive—meaning human powered—and dangerous workplaces. After locating an exposed face, men first had to pick and shovel the overburden, some of which was soil and some of which was inferior rock called "bastard stone." The men removed the overburden in wheelbarrows and carts, which gave them a ledge to work. To get good stone out, they cut a dozen ten- to twenty-foot-deep vertical holes with hand drills down into the wall of stone, which could take months; shoved in black powder; and blasted out massive blocks. A single explosion could provide a year's worth of work, if the block didn't shatter into too many pieces upon detonation.

By the end of the Civil War, machines had replaced manpower. First to arrive were the multibladed gang saws. They replaced the single-blade crosscut saw, which required two men to cut through a block, as if it were a piece of wood. The mechanical gang saws needed only one man, who fed water and sand that were ground into the blocks by the metal blades, enabling him to cut multiple slabs at one time.

Much more important to quarrymen was the development of the steam channeler, which looked and worked like a cross between a sewing machine and a small-scale locomotive. The cutting tool consisted of five steel bars, each tapered at one end to a sharp chisel edge. Steel bands held the bars together in a row, so that from the side the bar ends looked like fingers of a hand.[7] Steam power drove the sharpened bars into the limestone at up to 150 strokes per minute.[8] Creeping along a set of tracks the channeler took two days to chisel a 7-foot-deep by 50-foot-long by 1¾-inch-wide cut.[9]

Much safer than blasting, channelers also reduced waste and cost between one-sixth and one-eighth as much as hand tools. Channelers made men more efficient, too, with two men now able to do the work formerly done by twenty-five, which made the quarry owners happy.[10] They could pay the workers the same rate, get more work out of them, and make more money. But the unremitting pounding of steel on stone made men deaf, and the coal-powered channelers produced smoke "thicker than hell," as one quarryman recalled.

"Everything was done with hand signals in the quarries," said George Jones, general manager of Indiana Limestone Company (ILCO).[11] We were standing on the edge of the company's massive quarry north of the small town of Oolitic and watching men quarry stone about two hundred feet away. "I knew this one guy who had worked the Reed Quarry. He was in the hospital with cancer and couldn't talk. His family communicated with him by hand signals," said Jones. As we watched, he said that two of the guys below just told each other that they would turn the cut in eight minutes. They did.

Because removing the overburden was expensive and time consuming, quarries generally did not grow horizontally; why remove waste when you could cut down and use all the rock you cut? Vertical cutting, however, led to a small challenge. During channeling, quarrymen typically cut their 1¾-inch-wide grooves from one end of the quarry to the other and across in a perpendicular pattern, so that the ground looked like sliced, rectangular brownies in a pan. How then to remove the first piece, which would let the quarrymen access the surrounding material? Simply modify Mr. Tarbox's method and pound wedges into the channel on the long side of a block until it snapped at its base. This key block may not have broken cleanly, but the quarrymen could drill into the snapped block's upper edges, attach hooks, also known as dogs, and lift it out with the derrick, now also powered by steam.

Into the hole created by the removal of the key block, or more often key blocks, men took the next great power tool, the steam drill. They cut horizontal holes every six inches at the long base of an adjoining block, inserted more wedges, and pounded the wedges with a hammer until the block separated from the rock below. Hookers then attached dogs to the massive block, which could be four feet wide by seven feet high by thirty or forty feet long, and pulled, or turned, it onto its side with ropes attached to the derrick. Another set of guys split the now horizontal block into smaller pieces with more holes, more wedges, and more pounding. Finally, hookers attached the block to the derrick and everyone hoped the block wouldn't fall and crush them.

About the last process that required humans actually working with cutting tools was scabbling. Despite its success, Mr. Tarbox's method was not perfect; many blocks broke off with projecting bumps, edges, and knobs. The scabbler's job was akin to a plastic surgeon's: Remove the offending irregularities and leave behind a smooth and beautiful face. But scabblers' days couldn't last forever, and by 1907 a machine utilizing disks with steel teeth had replaced the stone surgeons.

With the rise of the machines and the spread of the railroad, Salem Limestone was on its way to becoming America's stone. In 1877 twenty quarries produced 339,153 cubic feet of stone, roughly enough material to adorn one and a half Empire State Buildings, the most famous Salem-covered structure. By 1895 forty-eight quarries produced 5,368,307 cubic feet and by 1912 the number had risen to 10,442,304 cubic feet. Sales peaked in 1928 at 14.4 million cubic feet, about 70 percent of all exterior stone sold in the United States that year.

Although saying this in front of a quarryman might not be prudent, modern quarrying looks rather boring compared to that of the past. Or as one quarry owner put it, "We were a lot tougher back then." Gone are the Eiffel Tower-esque derricks with their head-cracking cables, body-skewering hooks, and bone-crushing dropped blocks. No longer do coal-powered channelers pollute the air with noxious fumes or ruin ears with incessant pounding. And in the most up-to-date quarries, no one swings a hammer. Technology has taken over the least technological building material in the world.

Quarries now look like a young boy's dream with oversized dump trucks carrying blocks the size of an SUV and massive front-end loaders cruising around the yards lifting, tipping, and stacking stone. Front-end loaders have also allowed quarries to tidy up because the machines can more easily move and stack blocks anywhere they need to be. But the big rigs have created a new hazard at quarries, the potential to be run over by a truck. "It's the law of gross tonnage. Get out of their way," said Jones as we sat idling in his pickup waiting for a mountainous truck to pass.

Machines developed in the past few decades have further reduced the number of men working in the quarries. At the ILCO quarries, fewer than a dozen men worked on the stone. On the lowest level of the quarry, one guy drove a truck, two guys operated front-end loaders, and six guys worked on a recently turned block. Fifteen feet above them two men walked around the quarrying surface, which covered an area the size of a football field. Running the length of this surface were rows of crisscrossed channels. Several rows had been removed and the quarry resembled a half-eaten pan of limestone brownies.

In bygone times, a channeler would have made these cuts, but at this quarry, as at all other quarries in the Belt, a diamond belt saw cut these lines. It looks like a chain saw fashioned for Paul Bunyan, with a 16-foot-long blade extending out of a white box the size of two telephone booths. Instead of a steel cutting belt, the blade has a 1½-inch-wide polyurethane belt studded every 2 inches with raised metal plates. Microscopic diamonds encrust the stubby, T-shaped plates, which feel like sandpaper, gritty but not sharp. As the belt spins, the plates can cut through 16 feet of limestone at 4 inches per minute.

These are the cuts that Todd Thompson and Brian Keith like because when the blocks are removed they leave behind a smooth face. "Yeah, it's like kids in a candy factory when we get geologists in here," said one quarry owner, when asked if he worked with geologists.

The diamond belt saws require constant streams of water, which keeps the blade and belt cool while cutting. At the quarries water gushed out of the previously cut grooves and onto the quarry floor. Water also keeps the dust down. As with most quarries, water use requires the Indiana quarries to operate only in above-freezing temperatures.

In addition to using specifically designed equipment, ILCO workers also borrow technology. The two guys walking across the quarry surface crisscrossed with channels were using a product originally developed for rescue workers: industrial air bags. The workers took gray air bags, about 2 feet by 3 feet, and dropped them down into the 1¾-inch-wide channel closest to the exposed, vertical face of the limestone. Each bag was connected to an air compressor that slowly inflated the bags. As the bags grew, the men dropped additional bags into the widening gap, until the 130-ton block tipped over and onto a pile of rubble on the quarry floor. Quarrymen call this "turning a cut."

Men then clamored onto the rock with hydraulic air drills. Bringing Tarbox's technique into modern times, they drilled three twenty-inch-deep holes in a row and inserted a hydraulic expander, the arms of which spread and split the rock. "Our goal is to have no more guys swinging hammers," said Jones. "We are one of the few quarries to use these expanders. They are faster and safer and split the rock more cleanly."

Other quarries still use hammers. On a turned cut, men drill holes only six inches deep and six inches apart, insert plugs and feathers, and hammer them into the rock. Because the plug and feather technique creates short drill holes, the ends of these blocks look like a grin of perfect teeth. These quarries sound different than ILCO's. From a hundred yards away and above the sound of trucks, air compressors, and saws, it sounded like a chain gang working on a railroad as metal pinged metal when the men pounded the plugs.

The end product of both the hammer and the hydraulics is the block. At nonworking quarries, they clutter the ground in haphazard mounds, as if the children of Oliver Wendell Holmes's Dorchester Giant had wandered west and continued to play. These blocks are the abandoned dreams of earlier workers. The stone may have had a flaw, such as fossils or a crack, or the quarry owners ran out of money and left behind a mess they had planned to clean up.

Blocks also litter working quarries. They may form protective borders around active quarry ledges or sit in long-forsaken piles dotted with trees, but most spread across quarry yards in row upon orderly row. Indiana Limestone regularly has over twenty thousand blocks that cover an area one-third mile long by one-quarter mile wide.

In a further nod to technology, every block at the ILCO yard has a barcoded label telling the cutting date, grade, and color. Color is critical because the most valuable stone, buff, costs six dollars more per cubic foot than gray and twelve dollars more than the least expensive, variegated, a combination of buff and gray.[12] To access information on the blocks, ILCO installed a wireless network across the twenty-three-hundred-acre site, so the workers could read the labels and transmit information back to the main office. "The system has really helped except that we had nonquarry people tapping into our network, so we had to give access codes to everyone on staff," said Jones.

Blocks are the basic unit of the building industry. Out of them will come the raw materials, such as panels, sills, and keystones, to be assembled into courthouses, homes, and skyscrapers or, if an artist gets the stone, sculptures, monuments, and statues. All the blocks need is someone to cut them open.

Cutting takes place away from the quarry at a mill, also known as the fabrication or cutting facility. Here, too, diamonds play an essential role. No one, however, would court anyone with these diamonds. They neither sparkle nor shine. Nor are they rare, and most can barely be seen except with magnification. Cutting implements in use in the limestone industry contain diamonds ranging in size from three hundred to a thousand microns, no thicker than the proverbial one thin dime.

Nor are these diamonds natural. Industrial diamonds are made in labs that replicate the high-pressure (fifty thousand times what we feel on Earth), high-temperature (twenty-two hundred degrees Fahrenheit) environment that forms natural diamonds. First created in the 1950s, most industrial diamonds originally came from factories owned by General Electric and DeBeers. Both companies still produce synthetic diamonds, but so do the former Soviet bloc countries, as well as China and India.

Although mills in the Belt utilize the same basic technology, they take very different approaches. The ILCO mill is the most advanced, with many computer-programmed machines. It was eerily clean, as if the computers had taken over, with no cables, wires, or trash visible in the large indoor facility. Only a few workers were doing much physical labor. One man sat complacently at his computerized control panel. He had only to watch and make sure nothing went wrong. The machine slid the block forward, lowered the guide bar, engaged the belt, and cut the block. In contrast, at a much smaller mill, the guy sitting outside and running the wire saw monitored it vigilantly. He may have been motivated by the several men crowded around a nearby wire saw that they had partially dismantled because it had started to miscut. He did have one advantage over the indoor stone guy: The noise produced by the stone screaming under the outside saw was not nearly as loud as at ILCO's facility.[13]

The big limestone slabs next get cut with diamond-tipped circular saw blades. The majority of limestone slabs end their trip through the mill after a couple of passes through a circular saw, resulting in some sort of product—slab, step, cap, or quoin—with squared edges. Most jobs require a single blade, but to fashion multiple, same-sized blocks, such as for thin veneer or treads, some slabs get ripped by gangs of up to twenty-two blades.

"Operating the diamond saw was the easiest job," said Bob Thrasher, who worked in the mills in the 1950s.[14] "I could do it by ear, just listening to the stone until it was done, and then moving the block. I read the entire Bible, Old and New Testament, working on the diamond saw." Judging from the stacks of books, newspapers, and *New Yorkers* in his home, Thrasher appeared to have kept on reading, despite losing an eye in the mill, when a wire snapped and hit him in the face. "My father worked in the mills, too. He quit one afternoon when he was loading a railroad flatcar with his boss and the chock came out. The car started moving and split his boss in two. Dad never went back."

Three common denominators stood out at the mills: Everyone was male and had on a hard hat and safety glasses. Beyond these similarities, they wore jeans, overalls, or shorts; earplugs or industrial ear mufflers; sneakers or steel-toed boots. There were wiry little guys and beer-bellied ones. A few looked as if this was their first post-high-school job, whereas some of the men looked old enough to have grandkids in high school. But no women. They were only in the offices. No one came out overtly and said they wouldn't hire women, not that they would tell an outsider, but most questioned whether a woman could take the physical requirements of the jobs, either in a mill or in a quarry.

The reliance on new technology hasn't eliminated all the old tools. Some milling occurs in a world little changed from the industry's glory days of the early 1900s. It is a world that relies on steel to shape stone, a world full of dust. (Fortunately for the workers, limestone dust, which is made of the same material as a Tums tablet—calcium carbonate—is much less harmful than granite dust, which leads to silicosis, a lung disease that killed many Quincy quarrymen.)

For big jobs, the millmen turn to the planer, which employs a custom-cut sheet of steel to slide back and forth along the edge of a block and shave the stone. With a handful of passes, the blade transforms a rectangle of stone to a graceful scrolled molding. Planing such as this stands in stark contrast to the way granite is cut, because no blade, no matter what the material, could shave granite. It is too hard. But beneath the limestone, piles of shredded fossils accumulate, as if the rock is going back in time and returning to its prelithification state.

"Every project requires us to make a new blade for the planer," said Will Bybee, president of Bybee Stone.[15] "We have a separate shop that just cuts new tools for us." Bybee was standing in the company's immense mill shed, at the north end of the Belt, on a spot where quarrymen have worked continuously since the Civil War. Inside the shed, it felt like time had fast forwarded several months to winter. A dappling of white covered every surface and gave the air a foggy feel reminiscent of early January mornings in Seattle.

Although they weren't working on any columns, Bybee pointed out the lathe they use to cut them. It operates on the same stone-shaving principle except that the stone also moves, or more specifically spins, as the blade slides along. Bybee's lathe can fashion pieces up to nineteen feet long, from ten-foot-diameter columns to slender balusters.

Abandoning the hypnotic planer, Bybee continued on to the most tool-intensive part of the mill, where stone carvers produce pieces no longer based on a straight line. Their work is basically a pointillist drawing in reverse, with the carvers creating their image bit by removed bit. Following lines they had penciled onto the stone and tapping carbide-tipped points and chisels, the men at Bybee's hewed an acanthus leaf on a Corinthian column, sculpted a woman's lips on a pool table–sized panel, and shaped egg-and-dart molding.

Carvers require the most experience in the milling shed, with each man having completed seventy-five hundred hours working as a carving apprentice, preceded by six thousand hours as a stone cutter. Most also have three to five additional years working stone before graduating to cutter status. Carvers use some of the oldest tools on site, with many passed down from generation to generation. The workmen's benches—called bankers—at Bybee looked like a chef's kitchen with a spread of cutting tools, toothed, curved, flat, and pointed. Various hammers hung above the bankers, although several of the men carved with a pneumatic, or air, hammer, basically a power-driven chisel.

"The carvers have the most freedom in working stone. About nine out of ten projects have no work ticket and require interpretation in

3-D," said Bybee. Instead of the standard ticket, which includes precise measurements and drawings, the carvers may be given only a roughed-out sketch or a photograph of a piece they are restoring. The men then create drawings, mark out their blocks with pencil, and take up their tools. "Occasionally, someone doesn't give us enough information but basically we can re-create any shape needed," said Bybee.

And those shapes last for decades upon decades, whether it's a statue of comic-book hero Joe Palooka standing tall in Oolitic, row upon row of Corinthian columns adding prestige to government offices in Washington, D.C., garlands welcoming parishioners through the doors of St. Vincent de Paul Catholic Church in Los Angeles, or incised letters designating the Alaska State Capitol in Juneau. Go to any city and you can find similar Salem structures.

You will have to travel to Indiana, though, to find some of the more unusual and moving Salem carvings. They are the tombstones in cemeteries throughout the Belt carved to resemble tree stumps. At Green Hill Cemetery in Bedford, a six-foot-tall pair with interlocking broken branches memorializes Mammie Osborn Maddox and Alonzo Maddox. Stone flowers "sprout" from the base of her tree with ferns "growing" from the base of his. Nearby stands the tombstone of Hattie Wease, who died in 1912. Her tree stump rises from a stack of horizontal cut logs. Above her name are an ax and mallet, carved with precise detail into the stone.

Other stumps depict vines climbing the bark, a lamb at the base of a child's tomb, doves nesting on branches, or frogs hiding in foliage. Not purely decoration, each design has symbolic meaning. A broken branch represents a life cut short. A frog alludes to resurrection. Doves symbolize peace. These are shibboleths, codes that united individuals to a larger community. Even in death the residents of limestone country looked to stone to forge a common bond.

One of the most famous tombstones honors Louis Baker, a twenty-three-year-old Bedford stonemason, who died August 29, 1917, when lightning struck him at home. His co-workers sculpted an exact replica of how Baker left his banker. On the upper edge of a slanted stone slab, supposedly the piece Baker was working on, they carved his metal square. Below rest a broad, flat chisel, called a drove, and a stub-handled broom, one edge of which abuts a foot-long pitching tool. A wider chisel leans atop a hammer that just touches the sharpened end of Baker's point. Nearby is the apron he tossed onto his mallet. The slab sits on another slab, propped on a bench so perfect in detail of the wood that one of the boards "warps" and others have cracks where someone, perhaps Baker, had overtightened the bolts holding together the planks.

The bench reveals not only the qualities of Salem Limestone—ninety years of weathering have not erased the details of individual straws of the broom—but also the qualities of the men who worked the stone. To honor one of their own, the men of the Belt produced a monument that reflected gratification in working with simple tools, pride in their trade, and respect for their co-workers.

Some people say we should thank Mrs. O'Leary's cow for the popularity of Salem Limestone. Legend holds that her bovine kicked over a bucket that started the 1871 fire that forced Chicagoans to rebuild their city. As happened in most places that rebuilt after a fire, builders chose brick and stone for the job and within a year, several hundred buildings had "shot upwards like grass after warm spring rains," as wrote a Chicago journalist.[16]

The standard story line of historians and promoters of Salem Limestone is that the "buildings that suffered least from the fire were of limestone."

Tombstone of Louis Baker, Green Hill Cemetery, Bedford, Indiana.

Newspaper accounts from the time, however, report that during the fire, limestone "seemed as though [it] actually burned like wood."[17] Builders were so prejudiced against the local stone, most of which came from nearby Joliet and Lemont, that in the first thirty days after the fire, most ordered brick, from as far away as Philadelphia. In the year following, builders worked extensively with sandstone from southern Ohio.

Chicagoans in the 1870s, however, didn't let wholesale destruction and poor-quality stone get in the way of politics and corruption, and soon the local limestones returned to their former prominence, most famously when the Board of Alderman chose Lemont stone for the new Cook County Courthouse.[18] Promoted by its powerful supporters, the Lemont limestone stayed popular through the decade. A few builders did use Salem stone, but some of them, apparently not wanting anyone to know, stained their Salem rock to look like sandstone.

Recognizing that they needed to counter the influential Chicago stone interests, several Indiana quarrymen began to get creative publicizing their stone. One promoter distributed small pieces of Salem Limestone cut into paperweights and vases to architects and contractors. Another enterprising booster wrote, "This purity insures absolute integrity on exposures to the fumes of coal, while the perfect elasticity and flexibility of the mass render it invulnerable to the forces of cold and heat, air and moisture." Others claimed that the stone cleaned itself and that it had withstood the ice age "scarcely changed in any part."[19] If you don't have money and power, then stretching the truth works well, too.

The true qualities of the Salem—durability, accessibility, and ease of cutting—ultimately proved superior to the local limestones. Furthermore, the fast pace of reconstruction overwhelmed the Chicago area quarries to the benefit of the quarrymen of Indiana. By the mid-1880s, architects such as the high-profile firm of Sullivan and Adler had begun to use Salem regularly, most prominently on their Chicago Auditorium, built in 1887. Others followed, demand grew, limestone-laden trains bore north, and Salem buildings spread across the Windy City.

Chicago wasn't alone. The first train car of Salem had reached New York in 1879. In the same year, William K. Vanderbilt chose Indiana stone for the first of the great Vanderbilt mansions. Architect Richard Morris Hunt later selected Salem Limestone for Cornelius Vanderbilt's "The Breakers," in Newport, Rhode Island, William Astor's Manhattan mansion, and George Vanderbilt's Biltmore House, in Asheville, North Carolina. Putting today's megamansions to shame, Biltmore required nearly 10 million pounds of limestone. With New York leading the way, others followed, and railcars carried block after block to Philadelphia, Boston, Kansas City, and Minneapolis.

An additional and significant boost came in 1893 with the World's Columbian Exposition in Chicago, the fair that had hastened the end of the long reign of brownstone. Daniel Burnham's White City reestablished white as a building color and reinvigorated classical architecture. Combined with landscape architect Frederic Law Olmsted Sr.'s refined fairgrounds, Burnham's buildings helped popularize what became known as the City Beautiful movement. Cities around the country began to look at how they could refine green spaces and public buildings to create a more dignified and cohesive vision.

Both marble and limestone suited the demand for the cut and carved features that paid homage to the elegance of Greek and Roman buildings, but limestone was cheaper and more accessible than marble, though it was more expensive than the white-painted wood used at the World's Fair. With Salem Limestone you could get the look without paying the price of the more classically correct marble.

Few cities were more successful in adopting City Beautiful ideals than Washington, D.C. Numerous marble-clad buildings did rise, and had been rising in D.C. for decades, but around the turn of the century blocks of Salem began to arrive in the capital. The first all-Salem building went up in 1911 and the Hoosier stone soon became "the workhorse of the building stones of official Washington."[20] Like the nameless bureaucrats who tallied our taxes, protected our parks, and doled out political favors behind the Salem walls, the stone did its job with little fanfare. Salem quarrymen also benefited in 1915 from the passage of federal legislation that required post offices having gross receipts between sixty thousand and eight hundred thousand dollars to have facing of sandstone or limestone, which led to more than 750 Salem-skinned post offices.

Not everyone was pleased with the success of Salem. Senators from other stone-producing states complained that because the chair of the House Buildings and Grounds Committee, the assistant secretary in charge of the Treasury Department's Procurement Division, and the Senate majority leader all hailed from Indiana, these honorable men may have showed some favoritism. Minnesota Senator Thomas Schall went so far as to propose a resolution requiring all monuments and public buildings to be made from granite or marble, no matter the cost.

Schall didn't need to worry. Although two of the biggest contracts for

Salem rock, each of which totaled two thousand carloads of stone, came during the Depression, Indiana's stone output declined rapidly after its peak in 1929. During World War II, several hundred thousand cubic feet of Salem was used for the Pentagon, but by the 1950s quarries and mills had begun to close down, permanently. A central reason was the rise of modernism and its use of "glass and whatever," as one modern Salem promoter told me. Architects Walter Gropius, Mies van der Rohe, and Le Corbusier preached simplicity and lack of ornamentation; their buildings didn't need a material that could be carved and cut into flowery shapes.

Small orders trickled in, but for the next three decades the industry remained stagnant and depressed.[21] The OPEC oil crisis, however, was a turning point. People began to realize the importance of energy efficiency, that glass leaked like a sieve, and that stone required less energy to produce and once on a building was more efficient. The industry has been mostly steady since then.

The quarries of Indiana continue to extract stone, and the mills of Indiana continue to cut and carve Salem Limestone in all shapes and sizes. The men of the Belt are leaner and more technology driven than in the heyday of limestone in the early 1900s, but they still produce a classic building material. Their product isn't flashy. It's not colorful or sensuous. "The Salem Limestone doesn't deny it's a stone. You don't put up a building with limestone to express anything but permanence," says sculptor Dale Enochs.[22] "The Salem isn't a screaming, sexy material. It has humility. I liken it to Indiana. It's part of us." It is also part of America.

# 7

# Pop Rocks, Pilfered Fossils, and Phillips Petroleum—Colorado Petrified Wood

*Here is a building worthy of study . . . worthy of a visit or many visits to see and observe these marvelous trees . . . they stand to remind you of the ancient ages long ago, and to serve you with the modern fluid which is the vital touch of our ultra modern motor-age.*
—Joseph Walter Field, *Lamar Daily News*

TWO CARLOADS OF rocks arrived by train in Bartlesville, Oklahoma, in July 1939. From a distance they looked like nondescript stone, but up close one could detect something special. They weren't just rocks, they were dozens of pieces of petrified wood, so well preserved that one could count tree rings and see the fossilized bark, insect borings, and ancient knots. Most of the fossil logs were tan to white but some had been stained red to orange. The petrified wood stumps averaged three to four feet in length and thirty to forty inches in diameter.

Frank Phillips, Bartlesville's most famous citizen and founder of Phillips Petroleum, had purchased the petrified wood. It had come from a dry wash that cut across a windswept grassland about twenty-five miles south of Lamar, Colorado. The wash was a good place to find petrified wood because intermittent stream flows eroded the rocky walls of the sandstone drainage and exposed the harder, more erosion-resistant fossilized logs. In addition, the open, rolling terrain around the wash meant that trucks could easily reach the heavy pieces of petrified wood. This was critical because the longer pieces weighed over a ton.

Few, if any, people in the small town of Lamar knew that Phillips had commissioned the delivery to Bartlesville. The supposed buyer of the fossils was a man named B. R. Teverbaugh, who was working with Ted Lyon, Phillips Petroleum's division manager in Wichita, Kansas. Teverbaugh, in turn, had hired A. H. Matthews of Lamar to locate the petrified wood. On June 13, 1939, Teverbaugh, in a letter marked PERSONAL & CONFIDENTIAL, wrote to Phil Phillips, nephew of Frank Phillips, "The writer handled this through a friend we have in Lamar, Colorado, so that it will not be known who was desiring the petrified wood."

Teverbaugh further noted in his letter that he and Lyon had tracked down the land's tenant, but that the Federal Land Bank of Wichita, Kansas, now owned the mortgage. To increase the likelihood of getting the wood at a low price, Teverbaugh and Lyon told the bank that they were the ones who wanted the fossils.

Teverbaugh also traveled down to the property Matthews had found to take a look for himself. He thought that there weren't many good specimens but that they could get twenty to twenty-five tons of wood, which they offered to buy from the bank for thirty-five to fifty cents per ton. Anything higher and the two poor men wouldn't be "in the position to take" the wood, wrote Teverbaugh.

Better examples of petrified wood had been found in the area, Teverbaugh reported, but someone else was already using the material. That someone else was Bill Brown, who had recently built a small gas station in Lamar out of petrified wood.

Brown's station was the driving force for Phillips's pursuit of the petrified wood. He had found out about the building because Brown had leased it to Phillips Petroleum. Phillips then tried—again on the sly—to buy the building and move it to Bartlesville, but Brown found out the buyer's identity and jacked up the price, and Phillips, who was notoriously cheap, refused to buy. Instead, he decided to purchase his own petrified wood and ship it to Oklahoma, where he would build his own gas station.

Teverbaugh struck a deal for 48,045 pounds of petrified wood, paying $24.20, twice his initial offer. In early July he had the wood shipped by train to Bartlesville, more than 375 miles from the Lamar source. Phillips's parsimony surfaced again and the petrified wood sat unused at his country estate, known as Woolaroc, for decades until workers made it part of a sculpture in an outdoor pond. In contrast, Bill Brown's petrified wood building remained a landmark for years, and although it hasn't been a gas station for decades, it continues to attract and inspire visitors.

Brown's gas station stands at the north end of the main business street in Lamar, a community of eighty-three hundred in Colorado's southeast corner. The town is quiet and well kept and is more reminiscent of Kansas—flat, rural, and surrounded by treeless plains—than Colorado. Located along the historic Santa Fe Trail, Lamar has long been more a passing-through point than a destination.

Finding the station can be challenging, particularly when the building's present owner, who sells tires and used cars, parks three massive SUVs in front of it. It's a good thing he doesn't sell RVs, as even a medium-sized one would completely hide Brown's fifteen-by-thirty-five-foot structure. The best time to see it is early in the morning when the low angle of the sun makes the building with its crenellated roofline seem enchanting, as if it had been erected for a fantasy movie. No wonder Frank Phillips wanted to take it home with him.

Made from several hundred pieces of petrified wood, the station looks as if it had recently emerged from the earth. The dominant color is tan to dark brown, although black, milky white, and rusty red pieces also dot the structure. On one piece on the side of the building, the end of a five-inch-wide, blackened branch protrudes out from a trunk. Another log forms a Y where it has split into two branches, and below the front window is another stump, its rings a dark carnelian against a rusty tan background.

Brown used his biggest logs of petrified wood in front. Two three-foot-diameter pieces flank what had originally been the bay where cars entered for lubrication and washing. The larger of the two weighed over thirty-two hundred pounds, according to a newspaper story written when the station opened. Brown also took advantage of two big logs to inscribe them with his name. Other mammoth trunks stand on either side of the front window and at the ends of the building. Brown incorporated one crooked fossil log as a waterspout. Along the top he placed many pointed pieces, giving the station the look of a turreted castle. Most of the blocks on the north side are about the size of lunch boxes; Brown used bigger logs on the south end. Nearly every piece of exterior petrified wood is in an upright position, like a standing tree. The average piece is at least a foot across and two feet tall.

The few pieces of wood that aren't upright are the ones used inside the building, on what one early writer called the "oldest 'hardwood' floor in the world." Unfortunately, a dingy gray carpet scattered with fallen slabs of ceiling tile covers all of the stone floor except for where the front door opens. The interior doesn't look anything like it did during the early years of the gas station. The present owner chopped off the back twenty feet of the building, which was not made of petrified wood, and now uses what is left for storage. Two pianos, a baby carriage, broken chairs, a computer monitor, and cardboard boxes have replaced the snacks, tires, lubricants, bathrooms, and service bay of old.

Brown chose his best piece of petrified wood for the lintel above the front door. In the middle is the base of a two-inch-wide branch that looks as if it had recently been snapped off so it wouldn't thump you when you entered the door. To the right are a series of holes that must have been made by insects when the tree was still alive. Around the holes the stone is a pale tan to reddish color that I associate with recently downed wood where the bark has been peeled away. Below this barky region, the stone is gray and weathered looking. You have to reach up and touch the petrified log to make sure it's not real wood.

Walking around the building, you can see another notable feature. Between the petrified logs on its north side, the mortar is green and raised as if someone did a poor job of caulking the seams. Judging from the consistent style used throughout the building, it seems the entire structure must have had green mortar, but it has almost completely faded on the south and east sides. Coloring the mortar green was not Brown's idea, although a newspaper article written at the time implies that it was. A young man building the gas station for Brown thought the colored mortar would give the appearance of vines climbing the stone trees.

We know about this young man only because of the fame he achieved after leaving Lamar. His name was Bill Mitchell and he later went on to invent Cool Whip, Tang, and Jell-O. He also developed, by accident, one of the great candies of my youth, Pop Rocks. I still remember the rumors kids passed along of how other kids had eaten Pop Rocks, drunk a soda pop, and exploded. Mitchell's daughter said that her father spoke with fondness of the summer he worked on the station. "I think he thought it was a special time," she said.

Mitchell's story of working on Brown's building compounds one of the confusing aspects of the gas station's history. In a letter Mitchell sent to the Lamar newspaper in 1978, he wrote that he worked on the building in the summer of 1930 or 1931. Other articles, pamphlets, and Web sites state that it was built in 1932. The assessor's records show the gas station was constructed in 1933, as does one of Brown's stone-incised inscriptions. No notice of the building, however, appeared in the Lamar newspaper until an article dated September 20, 1935, heralded its imminent completion.

"There are so many features involved in this unusual filling station that it takes a few minutes to properly sense its unique position. An all wood building approved in the fire zone, because it cannot burn! An all wood building with scarcely a piece of wood in it . . . THE ONLY PETRIFIED WOOD FILLING STATION IN THE WORLD!" wrote Joseph Walter Field on the front page of the *Lamar Daily News*. "Can you imagine the thrill of realizing a lifelong ambition, a dream of years, materializing before your eyes? Well, this is the way Brown feels as he sees this unique building nearing completion."

After finishing the station and signing his lease with Phillips Petroleum, Brown leased the station to others, who ran it. Two of those lessees were brothers Gene and Blynne Smith. "Service was much more formal then. Gene always carried a tire gauge and chamois in his pocket," said Dorothy Smith, Gene's widow.[1] "When someone pulled up, he'd rush out, check the air in the tires, the oil, and water in the radiator. He also washed the windows."

No matter when the station was built, it was big news in Lamar. The town, like the rest of the country, was in the middle of the Great Depression. Lamar would soon achieve some notoriety because of photographs

William Brown's gas station, Lamar, Colorado.

of titanic storms of black dust raging across the county and out toward Kansas and Oklahoma, but its main claim to fame prior to the gas station was the infamous Fleagle Gang robbery of May 1928, which left four men dead following the holdup of Lamar's First National Bank. After a yearlong, nationwide manhunt, the police and FBI captured one of the gang members, using evidence from a fingerprint he left on a getaway car. The Fleagle Gang case is the first where the FBI used fingerprints to convict a criminal. All four members of the gang eventually died on the gallows.

For Brown, the station fit into his long-term role as a town booster. He had sponsored Lamar's first swimming pool, helped start a chapter of Rotary International, and co-owned the first radio station. He employed over a hundred people at his family's business, the Brown Lumber Company. The gas station was a logical next step in drawing attention to Lamar. The newspaper also reported that Brown had purchased 480 acres of land around the petrified forest where he obtained his building supplies and that he planned to develop the area as a tourist attraction.

No one, however, has located any record of Brown's purchase of the land. Eighty-one-year-old Lamar veterinarian Elmer Sniff e-mailed me that Brown stole the wood from private property. "My family made several Sunday afternoon trips to the 'Petrified Forest' in the 'Little Cedars,' about twenty-five miles south of Lamar," wrote Sniff. "Our only admonition was to never take any of the petrified wood and spoil the pristine nature of the area. However, not everyone felt this way." Several other Lamar natives confirmed that Brown had not necessarily acquired his petrified wood via traditional, legal methods. Ironically, Field wrote in his cover story about the building that Brown built it because of the "added threat that these [petrified remains] might be permanently lost to the region."

"Petrified wood is the fossil for the common man," says Kirk Johnson, a paleobotanist at the Denver Museum of Nature and Science.[2] In traveling around the American West, he has seen dozens and dozens of petrified logs in people's yards. "Some of the specimens weigh hundreds of pounds. I am amazed at the efforts people go to to haul this stuff back home," he says. "I think it's because they can relate to the wood as a living organism. They can recognize it as a petrified tree."

I observe people making the petrified wood–living tree connection often in my part-time job as an educator at the Burke Museum of Natural History and Culture in Seattle. As part of the teaching program, we give the students the opportunity to handle a variety of fossils. The only fossil that every child can identify is the piece of petrified wood. The parent chaperones also are able to recognize the fossil and they get more excited by a two-inch-diameter piece of petrified wood than by a four-inch-long *Tyrannosaurus rex* tooth. No one can walk out of a museum and immediately see a dinosaur, whereas a museum visitor can walk outside and immediately encounter a living tree that looks identical to the 100-million-year-old fossil tree she just encountered.

Recognition also applies in nonmuseum settings. Most fossils one encounters in the field don't look like a distinct life form (and definitely not like they do in a museum) or if they do look somewhat recognizable, few people can identify the specimen beyond calling it a bone or a shell. In contrast, when someone finds petrified wood in the field, they are often able to recognize it as a fossilized tree and not merely a fossilized something or another. They can have a distinct sense of discovery, knowing they have made a connection to the past. It can be a profound moment, particularly if they have never found a fossil before.

Petrified wood has another advantage over most other fossils. Drop that *T. rex* tooth and it might break. Drop a piece of petrified wood and it might hurt the floor or toe it hits but it will remain intact.

One of the most beautiful of fossils, petrified wood is displayed in photos in Frank J. Daniels's book *Petrified Wood: The World of Fossilized Wood, Cones, Ferns, and Cycads*. One strange specimen from Arizona has quartz-filled cracks that look like a star in a night sky. A green and blue stump from Oregon resembles a topographic map of rivers and forests, and a fiery red, yellow, orange, and blue log from Nevada seems to glow from within. On many of Daniels's examples, the perfectly preserved annual tree rings allow you to determine precisely how long the tree lived.

Unfortunately, people's fascination with petrified wood often motivates them to break the law. Rangers at Petrified Forest National Park in Arizona estimate that visitors steal twenty-four thousand pounds of petrified wood per year. Many scofflaws argue that they don't know they are breaking the law, despite the numerous signs telling them otherwise. Some think that if they don't pick up the petrified wood, it will erode and be lost to science—still others take a more modern approach and blame their parents, claiming that it is family tradition to collect petrified wood.[3]

About a mile from the exits, rangers erected signs warning that cars are subject to inspection before leaving the park. The warning has had limited success, as each month, park employees collect just tens of pounds of petrified wood near the signs.

The park visitor center has a display of what rangers call "conscience letters" from people who were not caught during car inspections but who later returned their purloined petrified wood. Many letters refer to a curse. People said they lost their dogs, got flat tires, or had a death in the family because they had taken a piece of petrified wood. The display has been effective; follow-up interviews by park employees show that 80 percent of people who read the letters said they wouldn't steal wood—not because it was illegal but because of the potential bad luck. One unexpected consequence, however, was that visitors who read the letters sometimes stole a piece so they could send it back with a "witty letter," in the hopes it would get posted in the park's display. The park receives about six hundred pounds of returned petrified wood per year. Clearly the curse is only so effective.

Fossils such as petrified wood have long intrigued people. "No greater objects of wonder have presented themselves to man's consideration than the fossils which from earliest times have been observed in different parts of the earth's crust," wrote Lester Ward in his *Sketch of Paleobotany*.[4] And yet few understood how fossils formed until at least the late 1800s. Prescientific ideas fall into three camps: by magic, by seed, and by flood. Many early natural philosophers did not consider fossils as the remains of once living beings. Instead fossils were *lusus naturae*, freaks of nature, which some latent planetary force (*vis plastica*) mystically created. One sixteenth-century writer, George Agricola, attributed petrified wood formation to a stone juice, *succus lapidescens*.

The second camp was only slightly less mystical. In 1699 Welsh naturalist Edward Lhuyd described "exhalations which are raised out of the sea" carrying fish-spawn that got caught in chinks in the ground and became fish fossils, which "have so much excited our admiration, and indeed baffled our reasoning."[5] Lhuyd may not have had a clue about how fossils formed, but he has recently received credit for the earliest description of a dinosaur fossil. He, however, thought the dinosaur tooth was one of his vapor-produced, faux-fish teeth.[6]

In the best case of a religious reformer masquerading as a scientist, Martin Luther popularized the idea that fossil plants and animals developed from Noah's flood. He wrote in his *Lectures on Genesis* in 1535, "I have no doubt that there are remains of the Flood, because where there are now mines, there are commonly found pieces of petrified wood."[7] Although Luther's diluvial theory took nearly a century to gain popularity, it influenced many scientists of the seventeenth and eighteenth centuries, particularly Swiss paleobotanist Johann Scheuchzer, who reported that he had found the fossilized remains of an eyewitness to Noah's big adventure. For those interested in seeing the infamous skeleton, named by Scheuchzer *Homo diluvii testis*, or human who witnessed the deluge, all you have to do is travel to Teylers Museum in the Netherlands, where you will learn that Scheuchzer's eyewitness was a giant salamander.

"This view [Luther's] may seem to us a poor substitute even for the worthless dreams which had to make way for it, but when philosophically viewed it will be seen that it was really a decided advance upon these," wrote Ward.[8] Diluvialists may have been completely wrong about the Flood, but at least they acknowledged that fossils developed from formerly living plants and animals that had turned to stone after the mud from the Flood covered those who missed the boat.

These early writers suffered because all of them lived before anyone had a remotely accurate age for Earth. An origin in Noah's flood made sense in their minds because the planet was only six thousand years old and the only widely accepted big event that could have killed and preserved so many plants and animals was the deluge. Not until the 1800s did fossil collectors consider a time on Earth before the Flood and before Adam and Eve. Aided by the work of Lyell and Darwin, paleontologists (the term first appeared around 1830) began to recognize that different plants and animals lived and died at different times in the past. By 1885, when Ward penned his *Sketch*, he could confidently reject the "puerile speculations" of old and rejoice in the "true advent of science."

After recognizing the great age of fossils, paleontologists turned to another hallowed question pondered over the millennia by fossil enthusiasts. How did petrified wood form? Agricola's idea of a stone juice penetrating rocks and producing something that looks like bone or wood may sound a bit too mystical but he was on the right track; petrified wood requires mineral-rich groundwater in order to form. First, the tree must get buried rapidly to prevent oxygen from reaching the wood. Otherwise, fungi consume the wood. Burial that leads to wood preservation may occur via volcanic ash, which can preserve trees upright, or via fluvially deposited mud, which can bury trees transported downriver.

Next, a petrifying agent must enter the wood. At least forty different minerals have petrified wood, but the most common by far is quartz, or silica. Because quartz resists chemical and physical breakdown better than most other minerals, any tree petrified by silica usually lasts longer than one preserved by calcite or pyrite, the next two most common minerals to petrify. Other trace amounts of elements may also invade the wood and provide color, such as iron (red), manganese (blue to purple), copper (green), and sulfur (yellow).[9]

A lack of oxygen does not mean a lack of life. Bacteria in the wood aid the process of silification by creating acidic conditions that lead to silica precipitating out of the water. The silica accumulates on and in the wood's cell walls, which act as a template for growth. Petrifaction continues as the woody walls deteriorate and more silica fills the voids, faithfully replicating the tree's internal structure. This process of permineralization commonly occurs in the fossilization of many organisms.

The change from wood to stone straddles a fine line between deterioration and accumulation. The wood has to break down slowly enough to allow silica to penetrate and replicate the plant's cell structure. If deterioration occurs too quickly, the template will disappear and no structure will remain. If it occurs too slowly, no voids will form for silica accumulation. "Organic templating," as geologists call this mode of preservation, contrasts with a common misconception that silica simply replaces wood molecule by molecule. In fact, most petrified wood still contains organic material, up to 5 percent by weight.[10]

Experimental research has shown that petrifaction can occur very quickly. A German physician wrote in the sixteenth century that he had made wood as hard as flint in three years by cooking it in beer and burying it in his cellar. Modern researchers have also placed alder wood in hot springs and within seven years silica had permeated the wood. Perhaps if Martin Luther had conjectured that Noah had floated on a sea of hot beer, then we might believe the German theologian's geologic observations.

Geologists reject a young age because of the instability of the silica that impregnates wood. Under a microscope, quartz can look either ordered and crystalline, known as agate or chalcedony, or amorphous and watery, known as opal. The majority of petrified wood under 65 million years old is opal, whereas older petrified wood is chalcedony. Researchers account for this difference because heat drives off water, consolidates the internal structure of silica, and converts opal to agate, a process that normally takes millions of years. Beauty comes to those who wait.

Bill Brown was not unusual in using nearby stone for building. A lack of money and lack of good means of transportation often drove the decision to go local. Driving across country in 1996 through western Kansas, I remember being struck by the sight of a sign welcoming us to Post Rock country. After wondering if we had entered some midwestern enclave of Yanni fans, I started noticing that fence posts along Interstate 70 were made from stone and not from wood. Turns out that in a land of few trees and many cattle, the best way to build a fence was to use the local limestone, a yellow-tan rock easily quarried from just below the surface. The post rock era ended in the 1920s, when farmers and ranchers began to use wood and steel, transported by railroads and cars.

In cities, one of the easiest ways to find older local stones is to look for big building blocks. Prior to the popularization of steel infrastructures in the 1890s, builders relied on stone for structural support, which required building blocks large enough to withstand the weight of tall structures. Style could also dictate the use of massive blocks, particularly during the era of Richardsonian Romanesque architecture and its emphasis on a natural, rough-hewn look for building materials. Architectural preference can complicate the picture during this period, the last quarter of the 1800s, but many who chose big blocks based on Richardsonian principles still had to use what was nearby because of shipping costs. For the most part large equals local.

I often take advantage of these early builders' reliance on local rock to impress my friends when I travel. All I have to do is find the oldest stone buildings and I can get an insight into the local geology and have fascinating stories to tell. (Of course, I could be misinterpreting my friends' comments, such as "That's great, David. Why don't you go do some more research on that and find us later?") Using this trick is particularly slick in urban settings, where most geology is long removed or long covered by development. For example, in downtown Seattle, which lacks any visible surface exposures of rock, most of the old stone buildings were constructed with sandstone, which points to sediment-rich rivers as a former dominant regional environment.

Others have used local buildings to learn more about local geology. In the 1890s, while searching for fossils about ten miles east of Medicine

Bow, Wyoming, paleontologist Walter Granger chanced upon the remains of a cabin. To Granger's surprise the sheepherder who built the cabin had used fossilized dinosaur bones for the little building's foundation. As he looked around, Granger realized that dinosaur bones covered the hillside. A year later he returned with a crew from the American Museum of Natural History (AMNH) and eventually dug up the bones of sixty-four dinosaurs, including *Allosaurus*, *Apatosaurus*, *Diplodocus*, and *Stegosaurus*. The AMNH men worked the Bone Cabin Quarry for six years and transported over 150,000 pounds of bone back to New York, making the hillside one of richest discoveries of dinosaurs ever.[11]

The Wyoming sheepherder and others who incorporated local stone did so for practical reasons. In the 1920s and 1930s, however, builders started to incorporate unusual local rock for a different purpose, to attract that new breed of American, the motor tourist. Even in small towns, stone no longer had to be local; railroads could transport exotic rock to within reach of anyone who desired it, but some business operators recognized that they could use their local stone to differentiate themselves from their competition.

For seventeen years, Thomas Boylan had collected dinosaur fossils from the hills around his home, about six miles due south of the Bone Cabin Quarry. He had intended to assemble them into a complete skeleton, but when he consulted local paleontologists, Boylan learned he had collected bones from a hodgepodge of species, "so I abandoned the idea and proceeded to use them the best way I could," he once told a reporter. In 1932 Boylan decided to draw attention to a gas station he ran by erecting a twenty-nine-foot by nineteen-foot building, which he built by cementing together the 5,796 bones he had scavenged. He referred to it as the Como Bluff Dinosaurium and the Building That Used to Walk. Boylan, and later his widow, ran the gas station and dinosaur cabin, which they converted into a museum, until the 1960s, when Interstate 80 replaced Highway 30 as a major thoroughfare through Wyoming.

Boylan and Brown also promoted their enterprises by printing postcards. Both attracted the attention of Robert Ripley, who featured the buildings in his *Believe It or Not* column. Brown's card shows his station with no pumps, just a well-dressed woman and man. Ripley mentioned the station in his column on December 17, 1935, under the odd caption The Petrified Wood House, Built Entirely of Wood Turned to Stone.[12]

Boylan's card, labeled Petrifications on U. S. Highway 30 Como Bluff,

Wyoming, shows that he spruced up the building by adding a sign describing Wonderful Wyoming the Dinosaur Graveyard. On the back, he described the "torpid reptiles" who "thrived in torrid heat" and called the building a "5796 page book of creational hieroglyphics." Ripley highlighted the cabin in his newspaper feature on April 26, 1938, as the "Oldest" Building in the World.

Brown's building made it back into Ripley's again on September 21, 1991. The new caption read, The World's Oldest Building. (From a geologic point of view, Brown's building is made from older fossils than Boylan's.) To be closer to the truth, Ripley would be better off by referring to the Morton liquor store building as the "Oldest" Liquor Store in the World.

These stone structures exemplify a change in America that had started slowly at the turn of the century and by the 1920s was rippling through the country. It was a change based on new technology and a new natural resource, fostered by the relief of winning a world war, and propelled by a rising stock market. The change was quintessentially American. Or as James Agee put it in *Fortune* magazine in 1934, "So God made the American restive. The American in turn and in due time got into the automobile and found it good."[13]

In 1910 only five hundred thousand people owned automobiles, in part because nearly all of the country's lanes, paths, and thoroughfares began as routes for animal-powered vehicles, which were superior to gasoline-powered vehicles at negotiating ruts, mud, and washouts. With the addition of new road surfaces oriented toward cars, such as asphalt and macadam (invented by Scotsman John McAdam), restive Americans were now able to go forth and drive, traveling on roadways with names such as the Lincoln Highway, Black and Yellow Trail, Old National Pike, and Lone Star Route.[14] By 1920 over 8 million people had registered their cars; by 1927,[15] a Model T cost $385; and by 1929, every state had established gas taxes to pay for roads.[16]

America had long been a nation of transients and travelers. We liked to move and when trains came along we hopped on them and went, but the automobile offered several significant advantages. Motorists controlled their own destiny and didn't have to rely on someone else's schedule or someone else's limited routes and limited stops. A family traveling in a car could stop where and when they wanted. Plus they traveled at a more human speed, not the blur-inducing, scenery-bypassing rate of a roaring locomotive. "More reliable and powerful than a horse, more personal and approachable than a train, the automobile seemed to restore a human scale to machinery that had been lost with the onset of the steel age," wrote Warren Belasco in *Americans on the Road*.[17]

"After the autoist had driven round and round for awhile, it became high time that people should catch on to the fact that as he rides there are a thousand and ten thousand little ways you can cash in on him en route," Agee wrote in *Fortune*. First and foremost in making money was selling gasoline. Pumps appeared anywhere one could put up a storage tank, such as hardware stores, general stores, and people's homes. Curbside pumps became ubiquitous taking their "place on the sidewalk with the mailbox, the streetlamp, and the fire hydrant."[18] By 1920 15,000 gasoline stations dotted the country and by 1930, 124,000 stations blanketed our gas guzzling land.[19] Station owners weren't just dealing a commodity. In the words of one enthusiastic spokesman, "it is the juice of the fountain of eternal youth that you are selling. It is health. It is comfort. It is success."[20]

"The gas station . . . is undoubtedly the most widespread type of commercial building in America," wrote Daniel Vieyra in *Fill 'er Up: An Architectural History of America's Gas Stations*. What had begun merely as a way to distribute gas evolved into a full-scale sales and service center. Owners sold oil, lubricants, tires, batteries, and accessories. They added lifts and pits for oil changes and repairs. Some owners also had separate rooms for car washing. Most had restrooms and many offered food, drink, and tobacco. They had become the one-stop service station so familiar to the American roadside.

As stations were transformed in the 1920s and 1930s, station design entered its golden age. Vieyra describes four recurring themes. Most elegant were what he calls the Respectful buildings, which grew out of the City Beautiful movement and fostered urban pride. Many resembled Greek temples with columns and pilasters. Others alluded to colonial designs, complete with cupolas and pedimented porticoes, whereas in the southwest, stations took on a more Spanish adobe appearance. All were supposed to inspire the motorist to stop in and consume.

The second category was Functional buildings, catering to a motorist's sense of efficiency. The classic was Texaco designer Walter Teague's white box. Clad in enamel, with three parallel green stripes wrapping around the building, clearly labeled service bays, a glass-enclosed office, and indoor restrooms, the Teague box was streamlined, orderly, and incredibly successful. All you have to do is go to your corner gas station and you will see how Teague's design morphed into the modern box.

Domestic buildings, Vieyra's third typology, satisfied those seeking a more familiar or rustic look. Who wouldn't want to pull up to a small cottage, either Tudor- or otherwise English-inspired, and purchase gas? The fuel *had* to be high-quality; the clean, folksy salesman emerging from his pastoral home wouldn't mislead anyone. Would he? That's certainly what the gas companies hoped buyers would believe.

Domestic, Functional, and Respectful buildings did share one common theme. Large companies designed them to promote a corporate image. They wanted travelers to know that when they saw a Texaco box in Tampa, Florida, or a Pure Oil cottage in Westerville, Ohio, they would get the same good service they got from their local Texaco box or Pure Oil cottage.

Some station owners eschewed corporate branding and endeavored to attract motorists with whimsy. They built stations out of old planes and modeled them after lighthouses and windmills. They made them look like tank cars, Brobdingnagian oil cans, and colossal shells. While some places showcased local themes, such as the World's Largest Redwood Tree Service Station in Ukiah, California, and a monumental Mammy with a giant, beehive skirt in Natchez, Mississippi, other locales turned to more exotic imagery. No pharaohs ever lived or died in Maine, but travelers in the 1920s and 1930s could buy gas from a series of pyramid-shaped stations. The same traveler could also go to Bardstown, Kentucky, and purchase fuel at a tepee, perhaps the first tepee ever built in that part of the world. Despite the cheesiness of the pyramids, one can argue that evoking Egyptians in Maine is better than reinforcing stereotypes in the south.

Vieyra labels this architectural style Fantastic and delineates its golden age as 1920 to 1935. Bill Brown may or may not have known of these other fantastic structures when he came up with the plans for his petrified wood station, but his little building was emblematic of the times. Motorists didn't know what to expect on the open road. They were adventurers, seeking new sights and new experiences. Buying gas at a petrified wood gas station was part of the adventure; it was what restive Americans wanted when they traveled.

Brown further benefited from Lamar's location, a little over 125 miles west of the hundredth meridian and less than thirty-five miles from the Kansas border. When travelers driving U.S. Route 50, often called America's Main Street, happened upon the petrified wood station, they knew they had left the moist East behind and crossed into the arid West. Hell, the wood out here was so dry it had turned to stone.

"I think they look like warriors crossing the plains," said Carolyn Peyton.[21] She was referring to the thirty-two-story-tall windmills arranged in parallel rows across open ranchland, twenty-five miles south of Lamar. One hundred and eight of the white towers, each with three 112-foot-long blades, have been erected on nearly twelve thousand acres of land, most of it owned by the Emick family. Peyton, who grew up in Lamar and arranged this little expedition, was standing with Greg and Val Emick, who have driven us out to look for petrified wood on their property.

It is perfect country for windmills, windy, gently rolling, and treeless. Prickly pear cactus, yucca, and sunflowers are the tallest plants growing amid weathered grasses. Not much moves out here besides the turbines, cows, and grasshoppers, and at least one tarantula.

"People occasionally try and steal the petrified wood. Recently someone tried to put these full rounds into his pickup," said Emick.[22] He was referring to three bathtub-sized petrified logs resting on the ground. "They crushed the back end of their truck, which is how we caught them," he added.

"Do you have time to see another log?" he asked. "It's a bit of a drive." Back in his V-10 pickup, he drove out a dirt road, through several gates, one of which I almost electrocuted myself on, crossed a dry, rocky wash, and finally abandoned the road to head cross country. He stopped at a mostly buried petrified tree, unseen until almost stepped on.

Brown, black, and white, the log had eroded like an onion, revealing ring after annual ring of growth. Emick walked off its length and determined that it ran aboveground for forty-five to fifty feet. Hundreds of smaller pieces had trickled down the hillside below it, looking as if the tree had recently died and decayed into pieces. One chunk the size of a baseball had a creamy, translucent texture, warmed by the midday sun. It weighed about twice as much as a baseball and its fossilized bark felt like a dried orange peel, in contrast to the smooth, layered growth rings. The nearly perfect fossilization of the ancient tree made the connection between the living and the long dead especially palpable.

Emick guessed that Bill Brown acquired his petrified wood from somewhere around this site. Frank Phillips purchased his twenty-four tons of petrified wood from property about a mile west of here and a 1931 booklet places the petrified forest on what became Emick land. That publication refers to one log that was a hundred feet long and six feet in diameter. Because the exact spot of Brown's acquisition of his building materials will never be known, it presents a geologic conundrum.

When did the trees live? That is the 50-million-year question. Did they grow 150 million years ago in a dinosaur-rich savannah or 120 million years ago in a hilly landscape of fast streams or 100 million years ago, as a sea began to invade North America? Part of the problem can be summed up by one geologist, who said about the area, "I haven't seen enough petrified wood of that quality to build a bird house, much less a gas station."

The question is interesting because that 50-million-year period witnessed one of the great events of evolution. The appearance of angiosperms, or flowering plants, was as dramatic as that of trees 250 million years earlier. Encased seeds allowed flowering plants to spread across the globe, filling niches never attempted before by their naked-seeded cousins, the gymnosperms. Angiosperms led to a burst of evolution, as pollinators and consumers evolved to feed on nectar, pollen, fruit, and

Close-up of petrified wood, William Brown's gas station, Lamar, Colorado.

seed. Flowers responded in kind, trying either to encourage or discourage visitors. We now reap this revolution in evolution: Almost every plant we consume is an angiosperm.

Angiosperms also filled the world with new aromas and new colors. Paleontologists suspect that some of the dinosaurs were colorful, because their modern descendants—birds—are, but greens and browns still dominated the palette 150 million years ago. With the evolution of flowers around 132 million years ago, the landscape brightened with hues rarely seen before on land.

No matter when Brown's trees once lived, water carried them some unknown distance, perhaps tens of miles but more likely much less. A few trees probably remained intact but the majority broke into smaller sections, as you would see along any modern river. After they settled, most likely on the inside of a bend where the river slowed or possibly in a logjam, fine-grained sediment quickly buried the downed timber. Groundwater rich in silica infiltrated the wood. The silica impregnated and replicated the tree's structure. More sediments piled atop the tree-rich beds of sand, converting it to a sandstone.

As with so many other aspects of the history and geology of the gas station, paleontologists cannot tell what species the trees originally were. The best guess is some type of conifer but nothing more specific than that. They could be an angiosperm such as a magnolia or cottonwood, but probably not; gymnosperms dominated the planet's flora long after flowering plants appeared. No one has studied the petrified wood well enough, however, to say for sure.

I like to think that Brown's trees lived after angiosperms evolved. They grew in a grove near a fast-flowing river. Perhaps a herd of herbivorous dinosaurs grazed nearby, eating some of those newfangled plants with their colorful flowers. A few insects buzzed around, also periodically checking out the flowers. Wind blew gently through the trees. Geologists can be romantics, too.

Although Frank Phillips couldn't buy Bill Brown's gas station and never had anyone build him a copy from the petrified wood he bought in July 1939, he did become the owner of a new petrified wood building. On November 28, 1939, Phillips turned sixty-six. His employees, who called him "Uncle Frank," celebrated their boss's big day with parties and a parade. Gifts arrived from around the country. They included a portrait, numerous items emblazoned with the Phillips corporate logo and the number 66, two very ugly lamps, and a small model of a gas station.

Bill Quinn, the Lamar agent for Phillips Petroleum, station managers Gene and Blynne Smith, and tire distributor Red Mathews built the twenty-four-by-thirty-inch replica of the Lamar building. It had taken them several weeks to assemble. They first built a plywood mock-up and then headed south to the petrified forest to collect buckets of petrified wood chips. After putting together the model, they added a sign that read. The Only Petrified Wood Station in the World, two pumps, several cars, and a smartly dressed mechanic. According to Quinn, no other present "thrilled him [Phillips] more than the miniature petrified wood service station."[23]

Uncle Frank put his little gas station on display at his ranch, Woolaroc. It sat out for many years until Woolaroc opened as a public museum. The model didn't survive long. The public liked it too much. Greasy little fingers picked at it, breaking off souvenirs and stealing them, says Woolaroc's director, Ken Meek. He doesn't know exactly when but thinks the model was thrown out sometime in the 1950s, after Frank Phillips died.

Bill Brown's petrified wood building remained a gas station until he died in 1957 and then it stood unused until 1962, when James Stagner converted it to an office for the Lamar Tire Service. Stagner still owns it but hasn't used the building as an office for many years. He said that people still stop by and want to see it. Most are respectful and just take a few pictures and leave, but some want more. "A couple of kids once drove up and asked for a crowbar so they could pry off a piece. We told them to hold on so we could get two crowbars and we would pry off a piece of chrome from their car," he said. "They left without any wood."

After talking with Stagner, taking photographs, and exploring the building, I got in my rental minivan and drove north. I stopped at a gas station about an hour later. It was a typical modern station found near highways across America. No one came out to help me. I paid for my gas with a credit card inserted at the pump and got back on the interstate. Bugs still covered my windshield.

# 8

# The Trouble with Michelangelo's Favorite Stone—Carrara Marble

*Clearly it is a miracle that a stone, formless in the beginning, could ever have been brought to the state of perfection which Nature habitually struggles to create in the flesh.*
—Vasari, *Lives of the Artists*

*Nothing the best of artists can conceive*
*but lies, potential, in a block of stone,*
*superfluous matter round it. The hand alone*
*secures it that has intelligence for guide.*
—Michelangelo

ONE MIGHT THINK that an architect named Stone would know better. One also might think that a company that depended upon geology for its fortunes would know better. They didn't. Like so many others before and after them, Standard Oil Company of Indiana and its chairman, John Swearingen, and architect Edward Durell Stone had been seduced. Not by money or sex, the typical agents of temptation, but by stone. In particular, they had succumbed to the allure of Carrara marble. They made a big mistake.

In an age when skyscrapers covered in glass and steel dominated architecture, Stone rejected the "hygienic austerity" of human-made materials. He favored the "classic purity of the all-white building" with its "accumulation of history," and no stone better met this requirement than an "ageless material," such as marble.[1] Taking a less aesthetic approach, Bonnie Swearingen, John Swearingen's wife, said, "We're going to bring [the marble] all the way from Italy to Chicago."[2] Standard Oil's new headquarters would be in her words a "mountain of marble," piercing the Windy City skyline with the world's tallest marble shaft.

Stone and Standard gave in to marble for a simple reason. No material has as glorious a pedigree. For over five thousand years, humans have relied on the dignity, prestige, and beauty of marble. Whether in sculpture or architecture, East or West, public or private, sacred or profane, small-scale or monumental, marble bestows qualities found in no other stone. Use of marble signals that the artist or architect has achieved a certain status in life and can afford the stone of Michelangelo, of the Parthenon, of the Taj Mahal.

Construction on Standard's headquarters began April 6, 1970. Engineers describe the design as tube construction, with a central core for elevators, mechanical shafts, and emergency stairs, surrounded by a structure of steel and concrete. On the exterior hung a curtain wall of 44,000 marble panels, which served little purpose other than an aesthetic one. Each panel weighed between 275 and 350 pounds and ran on average 50 inches high by 42 inches wide by 1¼ inches thick. Four stainless steel anchors bolted to the building held each panel in place. The anchors looked like four hands with bent fingers wedged into four grooves on the panel's top and bottom edges. Because the groove was wider, deeper, and longer than each anchor, the panels could move slightly. After hanging the panels, builders caulked the grooves around each slab, so that from the outside you could not see the anchors. Nor could water get behind the panels and weaken the anchor system.[3]

Standard Oil's employees moved into their 1,136-foot-tall, marble-encased headquarters in December 1972. The honeymoon between man and marble lasted only one year. On December 26, 1973, a 350-pound marble slab plummeted from the eighty-second story and walloped the Prudential Building across the street. A windstorm two months later dislodged another slab, which destroyed a car on the street. Neither incident injured anyone.

Responding to a new city ordinance, Standard, now known as Amoco, made its first detailed inspection of the slabs in 1979. Over two thousand panels had cracks and seven had bowed convexly one-half inch. Workers replaced the potbellied panels with new ones and repaired the cracks.

Amoco also hired Texas A&M professor John Logan as principal geology consultant on the building. His specialty was studying how rock responds to temperature and pressure.

"The panels were intimidating. Anything that failed could be catastrophic," said Logan, now a geology professor at the University of Oregon.[4] He immediately set up temperature probes around the building to monitor it twenty-four hours a day. Ever since Lord Rayleigh in 1933 had shown that marble weakened as it was heated, geologists had known that thermal cycling warped marble. Logan also removed several panels from the building, took them back to his lab in Texas, and baked the rocks at three different temperature cycles. In all the tests, panels suffered a rapid loss of strength, which then tapered to a gradual loss. He also observed that larger temperature differences in cycles led to more pronounced loss of strength.

Coincident with Logan's testing, employees continued to notice cracks and bowing appearing regularly in the panels. Inspectors in 1989 found almost 8,300 panels with cracks. On many, a slight tap with a hammer expanded the crack from a fine, vertical line to a crescent shape prone to failure. Worse, maximum panel curvature had increased to 1⅛ inch. In agreement with Logan's hypothesis that daily temperature changes weakened the panels, approximately 80 percent of those most warped panels occurred on the south and east sides of the building, the two areas that received the most direct sunlight. Tests further showed that an average panel had lost 40 percent of its original strength and that weaker ones had suffered a 75 percent decrease.

"Despite the numbers, there were no significant issues that panels would fail. They had stabilized after their initial loss of strength. I think there was no scientific reason to replace all of the panels, if Amoco practiced good maintenance," said Logan. But science was not the only concern and lawyers got involved, concerned that even one failed panel might fall and injure someone.

"When a person comes in and says forty-four thousand pieces of marble on your building have to be removed, you go 'whomp!' Well, after I picked myself up off the floor, we said there has got to be some other solution besides removing the marble," said Roger Hage, a vice president of Amoco, to a building conference in 1993.[5] Engineers initially thought workers could cut each panel in half vertically and horizontally and bolt the four newly created, quarter-size panels to the building, but when they realized they would have to drill over 700,000 holes and caulk 176,000 panels, Hage's team abandoned their resistance and began to look for the best way to replace the marble panels.

Amoco first secured every panel to the building using two white stainless steel straps per panel. After considering options ranging from replacement on an as-needed basis to full replacement with aluminum, Amoco decided to remove every marble panel and put up new two-inch-thick granite slabs. They came from a quarry in Mount Airy, North Carolina, where Logan had made a detailed geologic map to determine the best rock. He also took slabs and blocks back to his lab and heated, soaked, warped, and weighed them. On site, more technicians checked the panels before they were attached to the building. "They were probably the most tested granite panels in history," said Logan.

The recladding project took four years from 1988 to 1992. Contractors designed and built a canopy to protect the public at ground level, erected a monorail system at five levels for moving panels and people, and installed hoists to raise and lower the granite. They replaced a floor and a half of panels per week, finishing in November 1991. Caulking ninety miles of stone and dismantling all equipment lasted until July 1992. Big Amy, as the building had been nicknamed, remained open throughout the process. (Once known as Big Stan, for its connection to Standard Oil, it is now called the Aon Center, after its principal tenant, the Aon Corporation.)

After removing the panels, which weighed over six thousand tons, Amoco received more than two hundred requests to use the marble. Five hundred tons went to a small company that made clocks, awards, and trinkets. The items sold at the Amoco headquarters for between $150 and $250. Nearly forty-five hundred tons ended up as golf ball–sized landscaping materials at various Amoco refineries. The final one thousand tons, crushed into fist-sized pieces, went to a local university as a lining for a pond.

Like a nagging sore, problems with Amoco's marble didn't end with its dispersal. One year after Governors State University lined their pond, hundreds of stocked bluegill, carp, and crappie died in the shallow water, right in front of a popular patio. Initial press reports focused on whether the marble or adhesive on the panels was responsible, but a study later revealed that a bacterial infection killed the fish. The marble was finally off the hook.[6]

★ ★ ★

If you want someone to blame for Amoco's marble problems, you have to look no further than Michelangelo. Whether in sculpting or architecture, he exploited the brilliance and luminosity of marble as few have ever done. His work gave marble, particularly Carrara marble, the prestige that made it the material to use in corporate boardrooms, prestigious law firms, and rarefied social clubs. Marble good enough for Michelangelo had to be good enough for Standard Oil.

Michelangelo first became attracted to stone at a young age, or so he told his biographer Vasari. "Giorgio, if I have any intelligence at all, it has come from being born in the pure air of your native Arezzo, and also because I took the hammer and chisels with which I carve my figures from my wet-nurse's milk."[7] Like so many others, his first experiences were hammering his local rock. Born in 1475 in Caprese, about fifty miles from Florence, Michelangelo spent his early years in the quarrying village of Settignano, where the stone carvers, or *scarpellini*, worked a blue gray sandstone known as *pietra serena*.

Suckled on stone or not, Michelangelo did not start as a sculptor. His father apprenticed the youngster to the great fresco painter Domenico Ghirlandaio. Michelangelo soon surpassed his master, who recommended his protégé to Lorenzo de' Medici, the head of Florence's ruling family. It was in Lorenzo's garden that Michelangelo discovered marble, at the age of fifteen or sixteen. Lorenzo was a renowned patron of the arts, with a rich collection of Greek and Roman sculptures, plus a resident sculptor, Bertoldo di Giovanni, a student of Donatello, the greatest sculptor of the early to middle 1400s.

Again, Michelangelo's precocious ability aided his development. Vasari described how the young artist so impressed Lorenzo with his first effort in sculpting—a copy of a faun—that Lorenzo invited Michelangelo to move into his house, gave him fine clothes, and allowed him to sit at the family dining table. Perhaps apocryphal, the faun has never been found, but Michelangelo did begin to work regularly with marble in the Medici garden. He remained in Lorenzo's care for two years, until his patron's death in 1492.

Over the next dozen years, while living in Bologna, Florence, and Rome, Michelangelo completed as many as ten sculptures. These include the *Sleeping Cupid*, which he dirtied up and tried to pass off as a Roman antique, and his first surviving life-size piece, a fleshy, staggering *Bacchus*. He also traveled to the quarries in Carrara, seventy-five miles east and north of Florence, for the first time to find a piece of marble. Out of a brilliantly white, crack-free block he carved the sublimely holy Vatican *Pietà*. Next came his colossal *David*, hewn from a block of Carrara first quarried in 1464, and later dragged through the mud to Florence and worked by two other carvers, before sitting outside in Florence for over three decades.

What unites his work, particularly the *Pietà*, *David*, and later *Moses*, is that Michelangelo had transcended his medium. He had become an alchemist, turning stone into living beings. When you look at any of these great statues, it is hard not to think that you are looking at works carved of flesh and cloth. Every fold, every muscle, every feature is so realistic that you expect David or Moses or Mary to become animate and to tell of the great thoughts revealed in their faces. It feels as if they are present.

By finding the essential elements of humanity and transmitting them to stone, Michelangelo had in turn bestowed a sense of grace on marble. No one who has seen the *David* or the *Pietà* could ever look at the material again and not be reminded of refinement, of the ethereal spirit of humanity. And those qualities eventually became synonymous with the stone, whether the viewer had seen Michelangelo's work or not.

Everything about *David* is awe inspiring—his size, his location in the Galleria dell'Accademia in Florence, the feeling of reverence in the air around him—yet only a few yards away is the Gallery of the *Slaves*, with its five roughed-out statues. The figure in each appears to be wrenching himself out of his marble bounds. In all five you can see the process of how Michelangelo chiseled a man out of rock, of how he mixed precise and rough cuts, of how he revealed the textures and light within the stone. These pieces are not made of flesh. You know Michelangelo was working with rock. They are not sensuous. They have a density and a mass. They are grounded.

Michelangelo described his process of sculpting as the "art which operates by taking away." Other sculptors have written of releasing the spirit or story within the stone. I do not have the experience or knowledge to question artists' beliefs, but the unfinished blocks illustrate a profound link between man and stone, a link where a man recognized the strengths and weaknesses of stone and worked with them to create astounding works of art. Exploited to its fullest by Michelangelo, the bond between stone and humankind is central not only to sculpting but also to architecture.

Like Amoco, Michelangelo also suffered for his decision to use Carrara.

Instead of losing face and 70 to 80 million dollars, as Amoco did, Michelangelo almost died twice to get at the stone.[8] In December 1516 he convinced Pope Leo X and Cardinal Giulio de' Medici that they should let him design a new façade for the church of San Lorenzo in Florence. The façade would be "both architecturally and sculpturally, the mirror of all Italy," wrote Michelangelo to the cardinal's treasurer and liaison, Domenico Buoninsegni.[9] Michelangelo proposed a more audacious undertaking than anything he, or anyone since antiquity, had done. The last great all-marble building in Italy had been constructed in 203, and the entire façade of San Lorenzo would be marble, including a dozen monolithic columns.

Although Michelangelo never ended up carving any pieces of marble for San Lorenzo, he spent years arranging for stone to be transported from Carrara to Florence. Ultimately, dozens of different-sized blocks arrived, some of which he later may have used in the Medici Chapel. He also sold blocks, but most of his marble probably was stolen after he left Florence for good in 1534.

His initial task was to locate good stone. Michelangelo could have worked with a middleman, who would find, cut, and deliver marble, but he didn't trust the ones in Carrara and nearby Seravezza, two of the main towns of the marble district. The men cheated him. They didn't understand marble. They didn't even know how to quarry marble, or so he wrote Buoninsegni.[10] In order to ensure good rock during the years he worked on San Lorenzo, Michelangelo traveled to the quarries, or *cave*, at Carrara and Seravezza nineteen times and spent eighteeen months organizing and supervising an ever-changing group of helpers. At Seravezza, about ten miles south of Carrara, he also had to coordinate building and widening several miles of new road, part of which required men with picks to cut a route deep into the marble mountains.

After finding the right stone, Michelangelo hired a crew of quarrymen, or *cavatori*, and *scarpellini* to cut blocks out of the mountain. First, they cut a narrow trench, then they pounded in either iron or wood wedges and forced the stone to split into a clean face. This is the same method Mr. Tarbox would "invent" almost three hundred years later. It is also the method used as far back as the Third Dynasty (ca 2686–2613 BCE) in Egypt. About the only significant difference between the early Egyptians and Michelangelo's *cavatori* was the evolution of the wedges from copper (up to 1500 BCE) to bronze (up to the eighth or sixth century BCE) to iron.[11]

The quarries were also his mill; the notoriously penny-pinching Michelangelo wasn't about to pay to transport excess stone or for stone that might have hidden flaws. To shape the stone, which might become a smaller block, a roughed-out figure still encased in a block, or a column, the *scarpellini* referred to pages of Michelangelo's detailed drawings. One book of his drawings shows twenty-two different shapes, many of which required several exact copies. By August 1517 eight blocks and three figures, along with another half dozen cartloads of marble, were ready for transport.

And here the fun began. Not only did Michelangelo have to figure out how to move his unwieldy blocks, by land, by sea, and by river, but he had to pay exorbitant fees. In ancient Greece, for example, transporting stone had cost ten times what it cost to quarry, and the price doubled for every hundred miles it moved overland in Roman times. By Michelangelo's day, fees had dropped, but still constituted a major cost of working in marble. He couldn't avoid the middlemen who demanded money for everything from oxen rental to harbor fees to storage dues.

In moving stone, Michelangelo, and for that matter all movers of masses, had a simple goal: to resist the pull of gravity. Any time gravity led a block astray catastrophe struck. A block could slide too quickly down a slope and maim or kill. A heavily laden cart could sink into a road built across a swamp. A block could drop from a hoist and turn a boat into driftwood. To counter the adverse and untimely affects of gravity, Michelangelo relied on men and rope. Neither came easily. He wrote his brother that if the Carrarese "are not fools, they are knaves and rascals."[12] A crew walked off the job taking the hundred ducats he had paid them, and others quit in the middle of projects. The ropes, one of which was 422 feet long and weighed 566 pounds, could take days to arrive from Pisa, Florence, or Genoa. Michelangelo's detailed records show that rope accounted for 18 percent of the total transportation costs. He also had to borrow pulleys, buy wood for sleds, and order custom-made turnbuckles and iron rings.

With all the equipment ready, the men tied the milled marble to a hardwood sledge called a *lizza* and slid it down *lizza* paths, or *lizzatura*. The ones in Carrara, which haven't been used in decades, look more like ski slopes than ramps for lowering multi-ton blocks. The *lizza* traveled on greased or soaped poles laid like railroad ties. Rope wrapped around posts embedded along the *lizzatura* slowed the descent. As the block passed over a pair of poles, men picked up the poles and moved them around to the downslope side of the block. Rope men kept the rope taut around the posts until they ran out of material and had to move their arm-thick lines to the next post.

"It has been a bigger job than I anticipated to sling it [a column] down," Michelangelo wrote to a friend in August 1518. The column was the first to be quarried for San Lorenzo, and the first marble column quarried since Roman times. Michelangelo continued: "Some mistake was made in slinging it, and one man had his neck broken and died instantly, and it nearly cost me my life."[13] They had gotten the column to within thirty-five yards of the road.

Seven months later Michelangelo tried to move another column. His workmen had lowered it only a hundred feet when a metal ring broke and the column shattered. "After it broke we saw the utter rascality of it . . . the iron in it was no thicker than the back of a knife," he wrote in April 1519.[14] Again, Michelangelo and his assistants almost died.

Not deterred by his near-death experiences, Michelangelo willed his marble blocks and columns off the mountains to a road for their five- to eight-mile-long journey to the Ligurian Sea. "Conceive a channel of water running over a rocky bed, beset with great heaps of stone of all shapes and sizes, winding down the middle of this valley; and *that* being the road," wrote Charles Dickens of an 1844 visit to Carrara. Nothing had changed in five hundred years, he observed. The carts were clumsy, the mistreated oxen often died on the spot, as did their drivers, "crushed to death beneath the wheels."[15]

Dickens may have exaggerated, but Michelangelo's columns and blocks eventually reached the sea, where a boat would carry the stone thirty miles down the coast to Pisa. To get the marble on the boat, which Michelangelo had spent several months locating, required building a ramp, digging a trench to lower the boat, and dragging the marble up the ramp. As workers loaded one block with a three-legged hoist, an iron ring broke. The boat was not damaged, and no one died, but the breakage delayed that shipment by another week. In Pisa, the men used another hoist to unload marble into a storage yard, where it sat waiting for the winter rains to arrive to raise the Arno River.

"I am dying of vexation through my inability to do what I want to do . . . the Arno is completely dried up . . . On this account I am more disgruntled than any man on Earth," wrote Michelangelo.[16] Even he had to wait on the weather. Winter was also a fallow time for farm lands, which allowed Michelangelo to hire oxen. He needed them to pull barges loaded with stone fifty-five miles upriver to Signa, an impassable point on the Arno about ten miles from Florence. Depending upon weather and the recalcitrance of oxen, the trip took from one to four weeks. At Signa, the men unloaded stone onto oxen-drawn carts for the final one- or two-day trip into Florence.

The first blocks reached Signa in January 1519. By March sixteen shipments ferrying forty-nine blocks had arrived. Michelangelo's first of a dozen planned columns didn't make it to Florence until two years later. No other columns arrived. Several broke or never left the quarry and six reached Pisa, only to vanish to history.[17] Despite his fame, Michelangelo's disappearing columns did not lead to the famous phrase "He lost his marbles." Or maybe it did; thirteen months prior to the arrival of the lone column, Pope Leo X had canceled the San Lorenzo project. Michelangelo didn't go crazy, but he did write that he had been "ruined over the said work of San Lorenzo" and suffered an "enormous insult."[18] Additional marble blocks arrived throughout 1521. Michelangelo decided to use the stone elsewhere.

Michelangelo's labors are the travails of countless others who struggled to get stone out of the ground and transport it across land and water. Quarrying has been called the most conservative of all crafts because it changed little from its origins more than four thousand years ago to the late 1800s, when machine power replaced manpower.[19] We rightly marvel at the great works of architecture from the preindustrial world, extolling their design, their ingenuity in construction, and their durability. Perhaps we ought to marvel more that they got any stone to the sites.

Prior to Michelangelo, no one was more obsessed with marble than the ancient Romans. They traded for it, stole it, quarried it, and taxed it. Archaeologist Rudulfo Lanciani estimated that fifty thousand marble columns arrived at the Roman port of Ostia, of which nine thousand remained in the late 1800s. Another archaeologist has called marble the "sine qua non raw material" of the ancient Romans.[20]

Early Romans, like all other builders, began by using local rocks. As early as the fifth and sixth centuries BCE, quarries on the Palatine Hill provided soft, olive gray tuff, a volcanic rock also quarried from the surrounding Alban Hills and Monte Sabatini. Tuff and other nearby rocks, such as travertine and lava, remained popular throughout the republican era, but by the Republic's waning days in the first century BCE, marble had moved to the forefront of popularity in the city.

Not everyone approved of the change. Pliny the Elder, in his *Natural History*, discussed marble's early users, all wealthy politicians, and concluded that the fashion for marble was the "leading folly of the day." He deplored marble as an extravagant display of luxury and, in some of the earliest condemnation of habitat destruction, he decried the hewing down of mountains simply for use as delights for the imagination. "One cannot but feel ashamed of the men of ancient times," he wrote.

Pliny was a lone voice of opprobrium, railing against a time-tested formula. For the Romans of the late Republic and early imperial age, marble equaled luxury, luxury signified wealth, and wealth translated into power.[21] As classics scholar J. Clayton Fant has observed, marble was a particularly good symbol of wealth because it was expensive, imported, and unnecessary.

Archaeologists usually point to Greece as the reason Romans turned to marble. Completely conquered by the second century BCE, Greece and its buildings stood as shining examples to the Romans, sort of like the big brother or sister a younger sibling wants to emulate. Beginning with the Cycladic culture over three thousand years earlier, the people of the northeastern Mediterranean had utilized marble for art, religion, and architecture. The Greeks chose white marble for their greatest buildings because of the stone's luminosity and sparkle, proximity, and ease of cutting, at least relative to granite. When the Romans subjugated Greece, they started quarrying the Greeks' local stone from areas such as Mount Hymettus, near Athens, and Mount Pentelicus, location of the quarries for the Parthenon's stone.[22]

Imitation may be the highest form of flattery, but working your former opponent's quarries also signified power. We not only conquered you, we are taking your stones for our monuments and you cannot stop us. Lord Elgin and the marble panels he pilfered from the Parthenon get worse press, but the Romans liberated Greek statues two thousand years before the British. They repeated a similar takeover when Egypt fell under Augustus. The conquerors acquired vast new quarries, as well as an obelisk or ten.

As their empire grew, so did the Romans' desire for marble and other flashy stone. From Numidia they quarried butter yellow *giallo antico*. The island of Chios produced *portasanta*, a stone often compared to roast beef in color and texture; its name translates to "holy door," a reference to its use as doorjambs at St. Peter's and elsewhere. In Turkey the Romans acquired the brecciated purple, black, and white *africano*. A stone that resembles broken Oreo cookies, with the straightforward name *bianco e nero antico*, arrived from St. Girons, France. The Romans went to Egypt and took a purple igneous stone, *porfido* or *lapis porphyrites*, from which came the words porphyry and purple. Back in Greece, they raided Thessaly for *verde antico*, an emerald and white metamorphic stone, known as serpentine for its resemblance to snake skin. Closer to home the Romans quarried their lone indigenous polychromatic marble, *cottanello*, a white and brownish red, swirled stone.

Although the Romans considered these rocks to be marble, modern geologists do not. To them, a marble is metamorphosed limestone. To the Romans, who called marble *marmor*—from the Greek adjective *marmareos*, meaning shining or shimmering—marble referred to any hard rock suitable for sculpture or architecture. Such "honorary marbles," to borrow a term from archaeology, included granite, breccia (rocks composed of angular fragments), porphyry, and serpentine. Not that this expansive definition died out with the Romans. Go to any store selling architectural stone and you will find a plethora of non–metamorphosed limestone labeled as marble.

Despite the abundance of colorful stone the Romans did not stop using white marble. Seeking a local source, they found it in the foothills of the Apuan Alps, about two hundred miles north of Rome. The Romans knew the stone as Lunense marble, after the town of Luna, a port on the Ligurian Sea six miles from Carrara, originally celebrated for its cheese, wood, and wine. Later heralded as Carrara marble, after the town nestled at the foot of the Apuans, the creamy white rock has been quarried for well over two thousand years.

Good evidence, including tool marks and written sources, shows that the Romans began quarrying in the second century BCE. By the time of Augustus's rule, from 27 BCE to 14 CE, Lunense marble had achieved its sine qua non status in public buildings. (Ever disgruntled by other people's ostentatious tastes, Pliny criticized the first person to use solid Carrara columns in his home.) Recently, archaeologists have dug down through the layers of the ancient stone dumps, known as *ravaneti*, that cover the Apuans and found pieces of cut marble carbon-dated at 763 BCE.[23] The Etruscans inhabited the region at the time and probably collected marble they found on the ground instead of quarrying it.

Carrara achieved its status in part because the Emperor owned the quarries. His minions established an efficient management system, which employed a host of functionaries to supervise skilled workers and slaves, manage expenses, and market stone. Quarries operated as assembly lines, offering semifinished decorative and architectural elements, such as columns, tables, entablatures, and statue bases. Skilled craftsmen worked on specific elements, aided by men who understood the strengths and weaknesses of marble and advised how to work the stone.

Finished products traveled from Luna by boat down the coast to the Roman port of Ostia and twenty miles up the Tiber River to an area in Rome now known as the Marmorata.[24] In contrast to Michelangelo, the Romans were able to transport massive quantities of stone because they had established a widespread and well-run network to do so. Successful transport of marble benefited from its elite status: The emperor controlled the system, which supplied stone for his projects and for his wealthy friends. You can imagine that most suppliers would want to stay on the good side of the emperor.

As the closest source of high-quality white marble to Rome, Carrara dominated the market for the next two centuries, when politics and the silting in and closing down of Luna's harbor raised the prominence of eastern marbles. But for those two-hundred-plus years including and following Augustus's reign, Carrara gave Roman buildings their face. Wander the Forum in Rome and you cannot help but notice the gleaming columns, the arches of Constantine, Septimus Severus, and Titus, or the elegant Trajan's Column. Wherever Romans wanted to show off their wealth they built with Carrara marble. Most marble, though, except for columns and capitals, served little structural purpose in buildings, whether grand or more utilitarian. Instead, inexpensive materials, such as brick and concrete, provided the structure and thin slabs of marble provided the look of luxury. Augustus's famed boast of finding Rome a city of brick and leaving it a city of marble was a veiled reference to this use of Carrara.

With the fall of the Roman Empire in 476, Rome settled into decay that lasted for a thousand years. A few popes tried to reinvigorate the Eternal City, but limited monumental building occurred. When Pope Martin V arrived in Rome in 1420 he found it so "dilapidated and deserted that it bore hardly any resemblance to a city."[25] Martin started a building and renovation spree that led to new roads, new churches, and a grand hospital. By the time Michelangelo first visited Rome, in 1496, the city had become a center of construction.

Walking around the revitalized Rome, like architects and artists past and present, Michelangelo would have encountered the ruins from the great buildings of the Roman Empire. They, too, would become his teachers. The ancient structures taught him about the vocabulary and motifs of architecture. Their scale inspired him and gave him the ambition to attempt a feat not tried since the glory days of the empire, to work with columns made from a single stone, as opposed to the common practice of building columns with drums of stone. Although he rarely followed the classical style precisely, Michelangelo "invariably retained essential features from ancient models in order to force the observer to recollect the source," as James Ackerman wrote in his classic *The Architecture of Michelangelo*.[26]

In Rome, Michelangelo would also have seen scores of Greek and early Roman sculptures. Modern art historians have argued that sculptures such as the *Torso Belvedere*, *Lion Attacking a Horse*, and the *Cesi Juno*, the latter of which Michelangelo called "the most beautiful object in all Rome," influenced, motivated, and stimulated him.[27] They challenged his understanding of marble and showed him the stone's possibilities.

He further learned about materials and quality from the great buildings in Rome. He saw the marble, how the Romans splurged by using solid blocks of it in their finest buildings, and how marble bestowed stature and elegance. Rome gave Michelangelo "a spark for explosions of fancy," wrote Ackerman, explosions that helped seal his reputation and that of marble, especially of Carrara marble.[28]

I first saw Carrara and its quarries from a car window, while driving north out of Pisa. Marjorie and two friends, John and Terry, and I were about twenty-five miles or so from the green foothills of the Apuan Alps. Broken clouds created a pattern of shade and light punctuated by bright white splotches, which several guidebooks explained: "No, that isn't snow, it's marble." After an initial giddiness at seeing the Carrara, however, my excitement began to fizzle. Where were the charming old buildings? Quarries had been worked in these mountains more or less continuously for the past five centuries and I expected quaint structures made of the local stone. Instead, the road felt like many other industrial/shopping mall districts I had driven by, with billboards, stores selling marble gewgaws, and warehouses creating a monotonous blur of banality. The stone mills—each with stacks of marble blocks, pallets of sliced stone slabs, and massive cranes for ferrying the stone—had a certain charm, but they looked like any stone mill you can see in Indiana.

Closer to the mountains the wide, industrial street gave way to a confusing maze of narrow, often one-way, poorly marked roads. After circling around, getting lost, backtracking, and hoping we knew where we were, we crammed our little rental car into a spot along a lane about twice as wide as our car. A typical Carrarese building, a four-story, stucco-covered structure, stood about two feet from the car door. On the other side of the road, and twenty feet below us in a concrete-sided trench, ran the Torrente Carrione, Carrara's couple-of-inches-deep trickle of a stream.

Abandoning the car, the four of us wandered toward where we thought our bed and breakfast might be. The streets closed tighter, the sharply rising buildings turning them into canyons. Eventually they became too narrow for cars. The windows of many of the buildings were festooned with drying laundry and hanging planters. At our B&B, no one answered, so we walked into the town center, crossing a plaza with white marble paving stones, each about the size of a brick. Marble sidewalks and curbs lined most streets. Not every building was made from marble, but we had clearly entered a hub of the marble universe. Not that the locals respected their great stone; graffiti covered many marble walls and several marble statues.

I like to think that these same narrow streets had greeted Michelangelo when he reached Carrara 510 years earlier on his expedition to find stone for the *Pietà*. When we finally got into our B&B, the owner told us that Michelangelo had worked in this very building and had carved the marble on the first-floor landing. The owner was a very nice person and I am sure she thought she was telling the truth. Perhaps Michelangelo had carved it to pay for lodging; money often was in short supply for him in his younger days.

The next morning, John and I met with Paolo Conti, a geologist from the Center for Geotechnology at the University of Siena. Conti had graciously offered to take us up into the mountains to learn more about the quarries and the geology. We got in his car and drove three miles or so into the Apuans and parked in a lot next to the Ponti di Vara, a stately, five-arched, brick-and-marble bridge. Originally a route for the railroad that crisscrossed the quarries, the Ponti di Vara is barely wide enough for the hundreds of trucks that zip across it each day.

From the parking lot, the quarries glowed a blinding white in the sunlight as they crept halfway up Mount Maggiore, which rose three thousand feet above us. Across the road yellow signs pointed to several quarries, including the Roman-era Fantiscritti, supposedly the source for the marble doorposts of the Pantheon. Another sign read VISITA LA CAVA IN GALLERIA PIU' BELLA DEL MONDO, Visit the Most Beautiful Underground Quarry in the World.

Pulling out several wonderful and colorful geology maps, Conti showed us that we were at the base of the Miseglia valley, one of three quarry valleys around Carrara. To the north lay Torano and to the south Colonnata. Each cut back into the Apuans for several miles and each had quarries first opened originally by the Romans. No one knows how many quarries have pierced these mountains, but one estimate runs as high as 650. Conti's geologic maps from 2000 listed 187 quarries.

"We are standing here at the edge of this tectonic window of marble, surrounded by sediments," said Conti, as he pointed to an oval of blue and purple on a geologic map of Tuscany that he had spread on his car. Blue and purple indicated metamorphosed rocks, mostly marbles but also some schists. Butter and pumpkin colors showed limestones, sandstones, and shales. The Apuan Alps were the largest zone of blue—representing an area about ten miles wide by eighteen miles long—on the entire map.

The Carrara began life as one of those limestones 200 million years ago, about the same time the brownstones began to form. North America and Africa had begun to split apart from each other, opening the chasm that would grow into the Atlantic Ocean. Across the globe, another ocean, called Tethys, had begun to spread onto a shallow platform off the coast of what we now call Italy.

This young sea was calm, clear, and warm. A smattering of islands popped out of the water. Along the shore, tidal flats and lagoons shifted back and forth, depending upon sea level. Sandbars made of oolitic grains formed seaward of the lagoons and a fine-grained mud of calcite accumulated on the seafloor. The few animals that did live in the sea included ammonites, foraminiferas, and bivalves.[29] The Salem Limestone formed in a similar environment, only on a much larger scale and in a sea far richer in animal life.

The shallow sea remained for several million years until an arm of the newly forming Atlantic Ocean invaded. The new waters dropped the temperature and lowered the salinity, precipitating an environmental crisis that halted calcite production. Enough calcite, however, had accumulated in the Italian sea's short life to later lithify into an eight-hundred- to eleven-hundred-foot-thick sheet of limestone.[30]

Little of consequence happened to these limestones until 27 million years ago, said Conti, when a small tectonic block, called the Corsica-Sardinian Microplate, rammed into the Italian peninsula. Corsica-Sardinia carried ocean-derived basalts, gabbros, and sediments and plowed them in a southwest–northeast direction atop the 200-million-year-old Italian limestone. For 10 to 15 million years, the persistent little plate rammed its larger neighbor, slowly piling and stacking the oceanic rocks atop the limestone.[31] Under four to six miles of rock, the temperature in the limestone reached to 300 to 450 degrees Celsius and converted the limestone to marble.

During this process of heating and squeezing, the texture of the original calcite changed as the mineral grains became more stable and more tightly packed. Crystals interlocked with neighboring crystals because the compressive forces eliminated excess pore spaces. Recrystallization often creates large crystals of calcite, which allow light to infiltrate deeply into the stone. In addition, because calcite crystals have regular cleavage planes, light bounces off these weak layers and the stone glimmers like a jewel.

The limestone further changed on a macroscale, as it metamorphosed into marble. Ten million years of collision deformed the once horizontal limestone beds into a humped mound of folded rocks, which became more humped and more tightly folded over time. The mound, however, couldn't sustain the squeeze and like a failed soufflé became too steep, collapsed, and spread. At the same time, the Corsica-Sardinia Plate stopped colliding and began to retreat, leaving behind marble buried under a mountain range as high as the Rockies.

As Corsica-Sardinia started to pull away from Italy, it stretched and thinned the ground surface, like when you pull on either end of a piece of gum. Conti explained that the surface rocks responded initially by rising to form a dome and then by breaking into a series of parallel basins and ranges, one range of which we call the Apuan Alps. Later erosion removed the overlying material and revealed the Carrara marble, now exposed at the surface as a complicated mess of folded, fractured, stretched, and squeezed rock.

Not all marble forms from burial of limestone. The Yule marble, the stone of the Lincoln Memorial and the Tomb of the Unknowns, formed when granitic magma intruded into a body of limestone and baked it. Contact metamorphism generally happens very rapidly, on the order of one hundred to a thousand years, and on a much smaller scale. At the Yule quarry, in Marble, Colorado, you can walk directly across the contact from the granite to the marble, which runs on the surface for about two miles. In contrast, a regionally metamorphosed marble, such as Carrara, can cover tens or hundreds of square miles, one reason people have valued it for two thousand years.

To see more of the quarries, Conti suggested we head up above the next valley north, Torano. After driving for ten minutes or so, he swerved the car across the road on a hairpin turn and pulled off onto a very soft shoulder. His driving seemed like a typical geologist's, veering abruptly to see rocks, combined with an Italian's sanguinity at cutting across a blind turn.

"I often bring students here," he said, perhaps explaining his driving calm. "It's one of the better spots to see the thick beds of limestone that became the Carrara." We got out, I looked both ways, and we crossed the road to see the source rocks for the Carrara marble. Oaks and beeches, some of which had begun to change color, grew out of the gray, massive rock. I had encountered limestone like this before in many places. I called it "tearpants" limestone, in reference to its sharp, resistant edges. "We

Looking down into the Torano basin, Carrara, Italy.

haven't found many fossils in this rock but this is one place we have," said Conti. I looked but found nothing other than a few snails crawling across the broken edges of the lackluster, 200-million-year-old limestone.

Up and up we drove as the road climbed and wound steadily through the foothills. We passed through zones of pines and under a canopy of rust-colored beeches before stopping near a small lodge, where we hoped to find lunch. Since it was closed we walked across the road and hiked up a trail to the Refugio Carrara, one of the well-stocked huts that offer food and lodging throughout the Italian Alps. We did find lunch there and I got to accomplish one of my goals for the trip.

Over the past few years, a cured pig fat called *lardo di Colonnata* has achieved a certain status among epicures, but for Carrara's *cavatori*, *lardo* has been a staple of their diet for centuries—a cheap, abundant food that tasted cool and refreshing on a hot day. I knew I couldn't quarry stone, but at least I could eat like a quarryman. The Carrarese make *lardo* in their dank basements by curing raw pig fat in a tub of marble. Additional flavor comes from a combination of rock salt, pepper, garlic, and rosemary. Like the *cavatori*, I ate my thin slice of *lardo* with onion and tomato on bread. It had a creamy, translucent texture and melted deliciously in my mouth. I followed it with a shot of espresso. Geologizing doesn't get any better than this.

Energized by pig fat and caffeine, we headed back out to find rocks. Conti whisked us down the road to a spectacular view into the Torano basin, where I could finally get a sense of the scale of quarrying. In the center of the valley, fifteen hundred feet lower and three quarters of a mile away, a ledgy quarry, known as Polvaccio, stairstepped up the valley face. Polvaccio has been worked since Roman times and was where Michelangelo quarried his *Pietá* block. Through my binoculars, I counted eighteen ledges of marble, each of which, Conti explained, was between fifteen and thirty feet thick.

Road after road zigzagged up the nearly vertical faces, faces covered white in marble by decades of quarry debris. The bends on the quarry roads are so sharp that trucks cannot turn and instead back down every other switchback. More roads climbed the valley wall below, as well as the smaller valleys south and east of Polvaccio. At the high points of the southern and eastern valley ridgelines, quarries had lopped off the summits, creating openings shaped like gun sights. "I remember when there was a mountain there," said Conti. The view was one of the most spectacular and disturbing I have ever seen.

"A thousand trucks a day carry stone out of the mountains," he said, as he discussed how quarrying has changed in the past few decades. Most do not transport blocks of marble. More stone now leaves Carrara as basketball-sized hunks to be used as a powdered, industrial filler called ground calcium carbonate (GCC). You probably have used a GCC-enhanced product. GCC makes paper whiter and more opaque. It stiffens plastic, in products such as garden furniture and coffeemakers. Ground-up Carrara marble goes into paint to prevent corrosion. GCC is also a filler in toothpaste, so brush up—if Carrara was good enough for Michelangelo, it's good enough for your teeth.

As GCC quarrying has expanded in the past twenty years, it has come at a cost. GCC is made by blasting the marble mountains, collecting the smithereens, and crushing, grinding, and sorting. Although GCC produces less waste, or *ravaneti*, than quarrying for building or sculpting purposes, it generates a finer waste, the snowlike slopes of the Torano basin.

*Ravaneti* in the good old days consisted of fist-sized stones, impure or partially worked blocks, and perhaps a broken column or capital. Water from rainstorms could percolate down through the spaces in the loosely packed material. GCC *ravaneti*, in contrast, originates from finer-grained material that doesn't get caught in sieves. It forms impermeable layers within the older *ravaneti*. Water that formerly could soak into the ground now builds up on the steep slopes and what would have been insignificant rainstorms now weaken the *ravaneti* and trigger landslides. In 1996 and 1997 researchers recorded fifty-two slope failures, the largest of which slid over two thousand feet.[32]

GCC production has added another danger. Those thousand trucks have only a few roads to use through town and one goes by a school, which led a group of mothers to stage a protest by throwing garbage in the road and stopping the trucks. Conti explained that although the town owns the quarries, which are leased to quarry operators, marble so thoroughly dominates Carrara's economy that passing any regulations, either safety or environmentally oriented, faces stiff challenges.

Our final stop was in the Colonnata valley, at a small quarry where an earlier owner had placed a slab of marble carved with a quote from Dante's *Inferno*. Many Carrarese take pride that Dante spent time in the area in the 1300s and mentioned the marble in his epic poem.

Conti had stopped not to admire Dante's verses but to look at the quarry and the blocks perched on the ledges above it. One of

Michelangelo's assistants described the color of Carrara marble as being like the "moon reflected in a well." A lyric image, but Conti's and others' detailed studies show that Michelangelo's pure white variety, known as *Marmo statuario*, is rare. More often traces of carbon or pyrite shade the Carrara gray. The rock's location within the folds affects the mixture and texture of the impurities. Conti pointed out the most common types, blends of gray and white, called *Marmo ordinario* and *Marmo veneto*, as well as blocks of the rarer, pure gray *Marmo nuvolato*. Nearby were several blocks of the striped *Marmo zebrino,* which on one block had been tightly folded into a narrow V. Another block recalled a frozen sea breaking to pieces and is known as *Marmo arabesque*.[33]

Tectonic stresses also affected the internal crystalline structure. More deeply buried marble developed larger crystals of calcite with straight grain boundaries. This is the rock Michelangelo wanted because it is easy to work and allows light to penetrate deeply. In contrast, crystals in the heart of a fold developed an elongate fabric, like a grain of rice, a texture that resists weathering and bending but is hard to sculpt. Another rock difficult to sculpt but good for building panels resulted from Carrara getting slightly rebaked during its long-term burial and having its straight-edged grains changed to serrate-edged ones. These various textures can be found throughout the Carrara quarry district, with single quarries containing all three textures.

A detailed understanding of the internal structure has helped clarify one of the long-known curiosities of Carrara, the exploding block of marble. According to an American geologist I know, truck drivers would stop in a bar on the way out of Carrara and come back out to their rigs to find blocks that "just popped and exploded." Nothing so dramatic happened, said Conti, but the edges of blocks would start to fall apart because of the release of stress on strained, or deformed crystals. In essence the stone was stretching after being released from its millions of years of squeezing. Conti added that quarry workers have an easy method for dealing with potentially exploding stone. They sweep their work area clean each night; if they find debris on the ground in the morning, they know their quarry walls may not be safe.

John Logan, the geologist hired by Amoco to determine what went wrong with their Carrara panels, also has studied the Carrara's internal structure and geologic history. He placed much of the blame for panel warping on the stone's past.[34] First, that warm, shallow sea of 200 million years ago produced calcite that was the basis for marble formation.

Calcite has an unusual characteristic that emerges when the sun bakes marble on a regular basis. Thermal radiation starts to expand calcite crystals along one of their three internal axes, or dimensions. Heat-induced expansion is an attribute that all of us have experienced: Just think of a balloon sitting in a hot car.

Calcite also contracts along different internal axes when heated. Known as anisotropy, the change becomes a problem when calcite cools because crystals that have grown and shrunk jab into each other and cannot return to their original shape. Over time, with "thermal cycling" and the "mismatch in expansion and contraction," as Logan has written of the changes, the disturbed crystals can alter the shape of a panel.

Carrara's metamorphic history created the second problem for the Amoco panels, said Logan. Once anistrophic calcite growth weakened the marble, it became more susceptible to a release of stress, also seen with the "exploding" blocks of marble. Although some sections of the marble deposits experienced more folding than others, all marble in Carrara suffered during metamorphism, so all marble will release strain at some point in time. When that marble is a thin slab and it's tightly secured in place, it can warp and bow to the point of failure.

Near the Dante quote, Conti also pointed to several rusty pulleys mounted into the wall. They were left over from cutting technology developed in the late 1800s called the helicoidal wire saw, sort of an industrial-scale Rube Goldberg contraption. It worked like a modern diamond wire saw cutting down through the marble like a wire cheese slicer. But this cheese slicer was more akin to one that sliced cheese with a wire moving along at fifteen feet per second that started in the kitchen, traveled down a hallway, around a corner, and up a flight of stairs to the den, where the block of cheddar was placed.

The helicoidal wire consisted of three strands of steel braided into a quarter-inch-thick wire. Speeding along, the wire traveled away from a motorized drive unit, over wheels mounted on a guidepost or wall, to a movable rig consisting of four pulleys. By passing through the pulleys and turning ninety degrees at each, the wire formed a rectangle. *Cavatori* cut a block by lowering the four pulleys until the wire bottom of the rectangle slid across and abraded the marble. A slurry of sand and water facilitated the cutting. Helicoidal wires could be over a mile long to allow the steel strands to lose their heat. The wires dramatically improved quarrying techniques, but if a wire snapped, whipping across the quarry, the results could be calamitous and deadly.

Transport also improved in the nineteenth and twentieth centuries from what Michelangelo had to deal with. Trains arrived in 1876 to ferry stone to the sea, but not until the 1910s did tractors, called *ciabattoni*, or shufflers, replace the oxen-drawn carts used closer to the mountains. Another decade would pass before winch-pulled trucks resting on rails took over the human-powered *lizzatura* slopes. Instead of dying under the wheels of carts or getting crushed by blocks crashing down a *lizzatura*, the *cavatori* had the pleasure of breathing the dirty air produced by inefficient machinery.

None of these earlier advances showed up on Conti's quarry tour. All that remained were a few rusting pulleys and an abandoned *lizzatura* or two. Diamond wire saws, hydraulic hammers, trucks, and front-end loaders were the new tools of choice, just as they had been in Indiana and Minnesota. With the machines, the *cavatori* have cut deeply into the

Active quarry, Colonatta basin, Carrara, Italy.

mountains in the past fifty years and have removed more stone than in the previous two thousand.

As we left Colonnata and rejoined the main quarrying road back to Carrara, truck after truck rumbled out of the quarries. The majority were filled with blocks destined to be crushed and ground to powder. Only a few carried the big blocks needed for architecture or sculpting. Michelangelo's marble has now become more valuable as an industrial powder than as a sign of elegance and opulence. The fate of the Amoco marble, as a pond filler, is being replayed on a global scale.

"Everyone will tell you that Michelangelo's stone is the worst for building," said John Logan. "The rock is too porous." Michelangelo's *Marmo statuario* has a coarse-grained texture with smooth boundaries between grains, which allows water to seep into the marble and weaken it. Perhaps a greater irony than using Michelangelo's marble for toothpaste is that Michelangelo, who sealed Carrara's reputation for beauty and prestige, chose a variety of Carrara with such poor qualities for architecture.

Amoco's troubles exemplify how the prestige and allure of marble can lead architects and builders down a slippery trail, but Edward Durell Stone and John Swearingen were not alone. An international study in 2005 found marble panel warpage from Cuba to Australia, with a concentration in Scandinavia. Finlandia Hall in Helsinki, designed by the great architect Alvar Aalto, had its Carrara marble panels replaced in 1999 because the marble bowed concavely, as opposed to Amoco's, which bowed convexly. Finnish designers chose Carrara again. The new Carrara panels began to bow within a year, but this time they bowed convexly. Three towers in Lidingo, Sweden, are more peculiar; the marble panels bow convexly and concavely on alternate rows.

Like Stone and Michelangelo, the architects of these buildings had been seduced by marble. Each designer knew that when they used marble, people would read the stone and see it as shorthand for grandeur. They didn't have to boast or shout about their clients; the marble would speak for them.

"The trend of corporations [is] to recognize the value of good architecture and its influence on the morale and pride of its personnel and the prestige that architecture can give to a business enterprise . . . Apparently the belief that 'Good architecture is good business' is gaining ground," wrote Edward Durell Stone in his memoir, *The Evolution of an Architect*.[35] I wonder how the failure of the Carrara marble, particularly in a company that employed hundreds, if not thousands, of geologists, affected those employees.[36]

Perhaps good architecture might benefit from some good geology. Another Stone-designed, Carrara marble–clad skyscraper opened in Toronto in 1975. The panels of First Canadian Place have also bowed extensively, and on May 12, 2007, a 250-pound panel dropped fifty-two stories and crashed onto a roof below. First Canadian's owners haven't ruled out recladding the building.

# 9

# Reading, Writing, and Roofing—East Coast Slate

*He was like a general on a battlefield of slate.*
—Johnny Cash, "The Baron"

*Here was a demonstration that a slate could speak in a foreign tongue.*
—Hiram Bingham, *A Residence of Twenty-one Years in the Sandwich Islands*, 1847

I ATTENDED AN elementary school three blocks from where I lived. Built in 1906, it was an elegant three-floor structure made of wood on a foundation of brick. A portico with fluted Ionic columns framed the front entryway, although I generally entered through the plain back door because it was closer to home. Above the front doors, which repeat the pediment-column pattern of the portico, in black script, was written ISAAC I STEVENS SCHOOL. As every child who passed beneath those words learned, Stevens was Washington State's first territorial governor.

Wide stairways with banisters, perfect for sliding down, connected Stevens's three floors. The hallways had wood floors and dark wood paneling and were not well lit. In contrast, the thirteen-foot-high classrooms had windows that stretched almost to the ceilings. If you were lucky, you got to pull down the long window shades when the teacher hauled out the movie projector for some grainy black-and-white instructional film. At the back of each classroom was a small room where we hung our coats and stored our lunches.

But the focus of class was the slate blackboard at the front of each classroom. I am sure that one of the things each of my teachers did on the first day of school was to write her name on the blackboard in large, blocky letters. Miss Smith. Mrs. Bangs. Miss Baker. She wrote on a glistening blackboard, recently washed and ready for a new year of students. Along the bottom ran a wooden tray, which held erasers and fresh sticks of chalk. As the weeks of school progressed, the board turned gray, the chalk eroded and broke, and the tray became covered in chalk dust, but a quick wipe with a wet cloth could return the board to a pristine state and no matter how well worn the chalk was our teacher could always find some bit to write with.

Over the years, I probably went up to the blackboard tens or hundreds of times. That was where I struggled with spelling and where I discovered how to add, subtract, multiply, and divide. I am sure I got chalk dust on my clothes and in my hair. I am sure I occasionally wrote at the wrong angle and made the chalk screech as I went along. I know I wouldn't have done that, or run my fingernails down the slate board, on purpose, because I still cannot stand those agonizing squeals. I also remember that like Bart Simpson, I invariably had to write "I will not . . ." over and over again on a blackboard, a penance for some transgression or other.

When I wasn't bad, I got to go outside and clean the erasers by clapping them together or hitting them against the brick foundation. I remember that cleaning the erasers was a reward: I got to go outside, whack things, create clouds of dust, and do something to please the teacher. Other friends of mine, as well as my mom, remember eraser cleaning as punishment. I also thought wiping down the blackboard was fun. The surface dried from shiny black to flat black almost instantaneously, as if the slate abhorred being wet.

Friends from coast to coast told me similar stories of learning from blackboards, cringing at the chalk squeal, and cleaning erasers. Their association with blackboards continued through high school and into college. Stories abounded of legendary professors who would zing sleeping students with bits of chalk or equally infamous teachers who always had lines of chalk stains on their backs. Blackboards were as much a part of growing up as skinned knees, learning to drive a car, or dating.

Slate blackboards are a wonderful teaching tool. They don't break or warp. They can be cleaned forever, either with an eraser or with your hand. They produce a pleasing click-clack sound when written on properly. Often taking up an entire side of a room, they provide a blank wall for jotting down anything from music to drawings to numbers. They also seem eternal and permanent. Just think of the photographs of Einstein, or any number of mathematicians and physicists, writing out complicated equations on a blackboard and you will recognize the role they have played in education and communication.

Or consider how our use of slate blackboards has seeded our language. We wipe the slate clean. We chalk up something to experience. We refer to a tabula rasa, literally a scraped tablet, but more often defined as a clean slate. We vote for one of a slate of candidates. We are slated to do something and those who had a debt were formerly said to be on the slate. No other stone has contributed a comparable literary etymology.

The precursor to the blackboard was the school slate, a handheld tablet upon which students could write with chalk or slate pencils.[1] Made of slate with a wooden frame, school slates had been used for hundreds of years in Europe and were starting to become more widespread in America in the late 1700s and early 1800s.

Blackboards arrived in America with George Baron, a military academy teacher from England. According to historian Stephen Ambrose, Baron began teaching math to cadets at West Point on September 21, 1801, and "illustrated his lecture by making marks upon a standing slate with a white chalk, thereby introducing the blackboard to America."[2] By the 1820s blackboards had spread to primary schools and colleges in Maine, Connecticut, and Massachusetts.

Not everyone cottoned to the newfangled blackboard. In 1830, during a math class, the fine youth of Yale College rebelled against having to describe a theorem depicted on a blackboard instead of using the traditional method of merely reciting from a book. The "Conic Sections Rebellion" resulted in the expulsion of forty-three of the ninety-six members of the class of 1832. Not content with simply expelling the mathematic malcontents, Yale administrators sent the students' names to surrounding colleges alerting them to admit the reprobates at their own risk.

Teachers liked school slates because students could share and reuse them and teachers could bring them to exotic places. Some of the largest artifacts found at the Donner party's final camp in the Sierra Nevada are pieces of slate, thought by archaeologists to be school supplies carried across the country by Tamzene Donner. In 1820 missionary Hiram Bingham helped introduce writing to the Hawaiian islands with school slates.[3]

Not all early blackboards were made from slate. Many were simply wood boards painted with black paint, later modified to a specially formulated application called liquid slating. Toward the end of the 1800s, a wood pulp and cement mixture known as hypoplate appeared. These blackboards suffered either from cracking; chipping; warping; absorption of oil, dirt, and water; or uneven wear, but they were inexpensive, at least in the short run. For the long run, however, nothing could match slate.

Blackboard slate and the smaller school slates came primarily from Lehigh and Northampton counties in Pennsylvania. Smooth, durable, and uniform, the slate took chalk easily and legibly, didn't absorb water, and stayed straight and true. By 1905 the majority of blackboards sold in the United States were made of slate. Six years later, the *Cyclopedia of Education* reported on blackboards that "It is doubtless no exaggeration to say that [slate blackboards] . . . should be used [in] all brick, stone, or concrete buildings."[4]

Although I did not know it during my youthful days of scribbling on the blackboards at Stevens School, I could not have used a better combination of materials for writing, at least from a geologic point of view. If I could have looked at chalk dust under a high-powered microscope, I would have seen a skeletal menagerie of countless single-celled marine algae. Known as coccoliths, they lived by the billions in the upper surface of the sea and were less than twenty micrometers wide. Some had shells shaped like steering wheels, others like pig-snouts surrounded by ruffles, and many resembled a hollow disc with radiating fingers. When these plankton died they accumulated in a white ooze on the seafloor and later lithified into chalk.

Pennsylvania's slate also began as a muddy ooze deposited in an ocean, when rivers carried clay, silt, and sand out into a deep marine basin. The 450-million-year-old sediments first formed into shale, followed tens of millions of year later by metamorphosis to slate, under thousands of feet of rock. At present, up to seven thousand feet of slate beds make up the valleys and ridges around Pen Argyl, seventy-five miles north of Philadelphia.[5]

Chalk and slate are the peanut butter and jelly of the geology world. You can write on many surfaces with chalk, and other stones can write on slate, but no combination looks as good or works as well together as slate and chalk. It is a relationship that needs no special preparation or processing. Any chunk of chalk writes perfectly well on any slab of slate. No other pair of rocks has such an integral and elemental connection.

Slate has dozens of uses aside from blackboards, though none of the uses are grand or known for their exquisite beauty or elegance. Instead, slate became the most utilitarian of materials, the building stone you turned to because "it is non-absorbent and germ-proof . . . sanitary and easily cleaned," as one early promoter described it. Slate doesn't compress under a load so it works well as paving, floor tile, and steps. It also can be inscribed in great detail and resists erosion, making it ideal for grave markers. Hard and impermeable, it works well for desktops, wainscoting, and urinals.

We look to marble for its beauty. We turn to granite for its durability. We build with brownstone because the Vanderbilts did. Sometimes, though, we just need a rock that gets the job done. Nothing fancy or famous or beautiful. And for that, no rock surpasses slate.

If you had been born a hundred years ago, you would have rarely spent a day of your life without seeing slate. Your mother would have washed your clothes in a slate laundry tub, set you to rock in a cradle next to a slate-manteled fireplace, and chopped food on her slate countertop. Your father would have used slate if he worked at a leather factory, brewery, or printing shop. When friends visited, they could have tied their horse to a slate hitching post, stepped over a slate curb, walked over a slate sidewalk, and up slate steps to enter your slate-floored residence. As electricity in the home became widespread in the 1920s, your family would have purchased slate electrical panels, because they could be drilled cleanly, didn't warp, resisted fire, and didn't conduct electricity.

The years around the turn of the twentieth century were the halcyon days of slate, when it was the modern equivalent of plastic: ubiquitous and practical. You'd be less likely to encounter slate today, in a modern home. Slate floors and countertops have had a resurgence of popularity, but few people seem to want to go back to the slate laundry tub or slate refrigerator shelf.

Of all slate's many applications, roofing is by far the stone's most widespread use. In 1830 an estimated 50 percent of the homes in New York City and one-third of the homes in Baltimore had slate roofs. During the peak years of slate production in the early 1900s, 75 percent of all quarried slate went to roofing.[6]

No other natural roofing materials possess qualities as appropriate for roofing as slate. Fireproof, resistant to rain and snow, and able to withstand high winds, a slate roof doesn't rot or get eaten by insects. Slate shingles don't deteriorate in the sun. They don't curl and lose their water resistance with age. They also look good and can last for centuries.

Some of the earliest archaeological evidence for slate roofs comes from Roman-era England. The eighteen-by-twelve-inch shingles resemble the bottom end of a necktie, with a flat top and triangular bottom. A single iron nail held the slate in place. Although archaeologists suspect that the use of slate roofing faded when the Romans left, by the 1300s slate roofing was on the rebound.

Slate shingles kept the elements out of the great estates of the landed gentry and the cottages of peasants. Court rolls and manor accounts reveal that slate had become so popular by the fourteenth and fifteenth centuries that the more desperate stole cartloads of it, possibly for resale, but also for covering their own homes. Other, more industrious thieves attempted to excavate slate illegally, usually from land owned by the local manor owner.

By the time America's earliest colonists started to abandon England for the New World in the 1600s, slate quarrying and slate roofing were well established in Europe. No doubt most people who ventured across the Atlantic had seen a slate roof. Many may have known how to put on a slate roof and many more would have known of slate's fire resistance and durability.

In 1662, in what may be the earliest building code in the colonies, Virginia governor William Berkeley drafted "An act for building a towne." It stipulated that towns should be built with thirty-two houses, each made of brick two feet thick at the foundation, eighteen feet tall, and covered in a roof of slate or tile. At least some Jamestown residents complied; archaeologists have found roofing slate from buildings constructed in 1663. Seventeen years later, after a fire destroyed eighty buildings and seventy warehouses in Boston, the General Court enacted a similar resolution requiring slate or tile roofing.[7] Canadian statutes of the early 1700s also recommended the use of slate to cover buildings.

Officials forgot one little factor—few, if any, people could afford slate. Only a handful of slate-roofed houses are known prior to about 1750. They include Thomas Hancock's stone home, well-known for its early use of granite and brownstone, and the Slate Roof House in Philadelphia, built around 1690, which achieved notoriety for its singular use of slate.[8] In Canada about the only group who could afford the stone were the Jesuits, who clad a church, a college, and a convent in Quebec with French slate.

Slate for roofing in seventeenth- and eighteenth-century America generally came from Wales or England. Old World slate had several advantages over the domestic supply. Welsh slate cost the same or less and builders thought it looked better and was of higher quality. Transporting Welsh slate was also easier because it could be shipped from seaport to seaport, whereas a lack of trains or canals often made transport of American slate challenging.

According to slate-roofing historians, the first commercial slate quarry in the country opened about sixty miles east of Philadelphia, near the towns of Delta and Peach Bottom, Pennsylvania. A Welshman discovered slate there in the 1730s, but operations didn't start officially until 1785. To honor this quarry and its history, the Maryland Historical Society has erected a sign just south of Delta, on the Maryland side of the Mason-Dixon Line, highlighting this eminent fact. The sign also reports that Peach Bottom Slate was judged "Best in the World" at the "London Crystal Palace Exposition of 1850." Never mind that the London exposition took place in 1851 and the only slates to win an award in London in 1851 came from Wales and Sardinia.[9]

After the initial Pennsylvania discovery, as well as ones in Virginia, New York, Vermont, and Maine, the American quarries lumbered along with little growth. Not until the 1840s did the slate industry start to grow. Slate historian and consultant Jeff Levine has cited three factors in the switch from Welsh slate to American slate. First was the spread of railroads, which lowered the cost of shipping. Second was the introduction of architectural pattern books by people such as Calvert Vaux and Andrew Jackson Downing, who advocated ornate roofs as a design element and stressed the beauty and colorfulness of slate. Vaux in particular promoted the use of American slates over Welsh.

Levine's research into the history of the U.S. slate industry showed that the most significant factor was Welsh immigrants, who became the backbone of the industry. They began to arrive in large numbers in the 1840s, driven by poor working conditions, poor pay, strikes, and food shortages in their home country. The Welsh immigrated to all of the American quarry regions, which is why Pennsylvania, Maine, and Virginia have towns with Welsh-derived names such as Pen Argyl, Bangor, Arvonia, and Bethesda.

Slate quarrying, like all other stone quarrying in the 1800s, was labor intensive and dangerous, but it required a new twist, one related directly to the geology of slate. All slate quarried on the East Coast originated geologically as shale that later metamorphosed into slate. Central to the story was the Iapetus Ocean, which spread east from North America around 550 million years ago. Adjacent to the continent, the shallow water teemed with life, which led to deposition of limestone. Farther east lay a deep marine basin, out of which rose a volcanic arc. Fine-grained mud washed off the mountains out into the basin through submarine canyons, periodically interrupted by earthquakes that sent coarser sediments into the water and deposited beds of sand atop the mud. Deposition occurred slowly, on the order of one inch every three thousand years, and lasted from about 540 to 420 million years ago.

Environmental conditions did not remain stable throughout the millions of years of deposition. During warmer periods, sea levels rose and the water became stagnant. Little oxygen reached the deep sediments, which slowed decomposition of organic matter and facilitated the accumulation of material such as plankton. Rich in carbon, these sediments turned gray to black. During cooler periods, ocean circulation improved. More oxygen mixed into oceanic waters and lightly oxidized the iron in the sediments, turning them green and purple. Animals burrowed into and churned up the mud, leaving behind well-stirred, homogenous beds with few depositional features.

Nor did the volcanic arc remain stationary. Plate movement carried the islands toward North America. As the arc approached it created a bulge in the seafloor, analogous to what happens when you push a carpet into a wall. Sediment deposited on the bulge accumulated in an oxygen-rich environment, which turned the mud brick red.

Not all of the slates along the eastern seaboard followed this exact plotline. Red slate occurs only along the New York–Vermont border, where the best beds of green and purple slate also are located.[10] Slate quarried in Pennsylvania, Virginia, and Maine is almost exclusively black or gray, but within the thousands of feet of these sediments further variations exist.[11] Thin bands of sand, called ribbons, appear periodically. Some layers are richer in quartz, which makes the layers harder and less suitable to use for blackboards. Others have richer accumulations of iron-bearing minerals, which can weather and change color after the stone has been quarried. (Slaters refer to these varieties as "fading" or "weathering.")

To change to slate, the shale had to pass through the geologic equivalent of a trash compactor, getting squeezed and compressed. The walls of the trash compactor were the North American continent and a series of volcanic island arcs. As the arcs pushed into the beds of shale, the sedimentary beds began to fold, like when you hold either end of a piece of paper and move your hands together. Arc collisions happened 450 million, 410 million, and 370 million years ago. Squeezing not only compressed the rock by folding, it also drove out excess space filled by water in beds of shale. The new metamorphosed rock, slate, was harder and more dense than the original, unmetamorphosed shale.

Again, not all of the slates experienced the same level of metamorphism. The aforementioned "Best in the World" Peach Bottom Slate is harder than other slates because it was metamorphosed twice and at lower pressure and temperature, sort of like what happens when you slow cook bread and it becomes dense and tough. Its hardness led to its downfall because it cost too much to cut and shape. Quarriers determined that the best way to sell Peach Bottom slate was to grind it and use the granules on asphalt roofing shingles. You can easily determine hardness by wrapping a slate shingle with your knuckle; higher quality shingles ring instead of thud.

Squeezing also generated two additional and critical features of slate. Most important is the alignment of flat minerals such as mica within the beds. As the vise flattened the beds of shale, micas that were askew to each other began to rotate and align, perpendicular to the direction of squeezing. In addition, the heat and pressure from metamorphism dissolved parts of minerals, so that the grains' longest axes now ran perpendicular to compression. Think of mineral realignment as creating a rock with an internal structure akin to a deck of cards. Geologists refer to this alignment of minerals as slaty cleavage.

Cleavage makes slate an unusual building stone. For other sedimentary rocks, such as limestone and sandstone, the critical factor is bedding, the layers of sediments that formed during deposition, because quarrymen exploit bedding to split or cut stone. In contrast, quarrymen rely on cleavage for giving slate its principal quality, the ability to be split into sheets. Cleavage also gives slate its name; slate is a corruption of the French word *esclater*, to split. No other rock relies on cleavage for splitting.

Bedding and cleavage affect the quarrying of rock in a similar way: Quarrymen exploit a zone of weakness to fashion a block of stone. The central difference between bedding and cleavage is that cleavage generates an almost unlimited number of planes with which to split rock, because the split occurs between aligned minerals, whereas with bedding planes the split occurs between layers, which can vary greatly in thickness. If the bed is four inches thick, for example, it is hard to split the rock into two-inch layers. A quarryman can make a two-inch-thick layer but it requires cutting, a more machine-intensive process than splitting.

Bedding and cleavage reflect a fundamental geologic difference. Bedding is a first-order process. It develops during the original deposition of the rock. Cleavage is a second-order process. It develops after the rock originally formed, during a subsequent change induced by pressure and temperature. Metamorphosis also generates a secondary mineral alignment that affects quarrying.

During folding, elongate minerals such as quartz get reoriented and become aligned in the direction of the tectonic push, comparable to what happens if you slide your hand into a pile of toothpicks and they get pushed and turned parallel to your fingers. Known as sculp, this alignment gives the quarrymen a plane of weakness running roughly perpendicular to cleavage, as well as at an angle to the bedding planes of slate.

When the Welsh arrived in America in the 1840s, they knew better than anyone how to take advantage of cleavage, sculp, and bedding to work slate. They knew that in tightly folded rock such as slate, they needed to find a good bed, called a clear run, and follow it. Because the tectonic trash compactor pushed horizontally, beds folded vertically and clear runs often dove steeply into the ground, resulting in the narrow, vertical holes that characterize slate quarrying. To reach the rock, the Welsh introduced a hoist and trolley system that would allow deeper penetration than the prevailing dredge and crane. A trolley with a hoist ran along thick steel cables stretched between derricks on either side of the quarry. Men would be lowered into the vertical quarry on a wooden platform. The deepest known quarry plunged nine hundred feet into the ground at Pen Argyl, Pennsylvania. An apocryphal story goes that a quarryman once said that the pit was so deep that "you could see stars during the day from the bottom."

Once down in the hole, the men had three planes of weakness to exploit. They used a plug and feather system to break slate along sculp. They also could plug and feather along bedding. Sculp and bedding basically defined the length and width of a block. After forming these two faces, quarrymen cleaved the slate into a block by driving wedges into the stone. Whacking and wedging, they would raise the block just enough to slide a chain under it. Finally, they would lift out the block with the trolley and hoist that had lowered them into the quarry.

Once out of the quarry, slate was sculped and split by hand. After a block had been cut to a manageable size, a sculper took his block of slate and chiseled a perpendicular notch into one end, in the direction of the grain. He then placed a blade, called a sculping chisel, into the notch and hammered it with a mallet. Several hits later, a fissure propagated down grain and the block cracked into two pieces, which advanced to the splitter. Tilting his slab of slate vertically, so that it looked like he was facing the short end of a book, the splitter placed a broad thin chisel parallel to the top and bottom sides and tapped it with a mallet. He continued to split each block in half along a cleavage plane until he formed sheets of the necessary thickness, which has generally shrunk from about one inch to a quarter inch for roofing shingles. A third man trimmed the shingle with what looked like a dull paper cutter, notching and cutting, notching and cutting. Depending upon its size and where the trimmer worked, a shingle could be described as a "marchioness," "double double double doubles," "mumfat," or "Rogue-why-winkest-thou."[12]

He finished making a slate shingle by poking two holes into it with a treadle-powered hole puncher. To this day, sculping and splitting are crafts still done by hand. Attempts have been made to mechanize but nothing has proved superior.

To facilitate their work, sculpers, splitters, and trimmers kept the slate wet throughout the entire process. Like sandstone, coquina, and limestone, slate hardens and becomes unsplittable as the quarry sap dries out. Workers also had to worry about slate in the winter because ice splits the stone. In England, slaters described this process as the "hammer of frost."

In addition to the costs incurred by hand splitting and sculping, slate quarrying suffered from waste, which ran as high as 90 percent in some quarries. Many slate beds had too much sand, which made them unusable, but they were still in the way and had to be removed. Imperfections such as sandy ribbons or silica-rich concentrations, called knots, also marred good slate beds. And men made mistakes, lots of them. At some quarries, the massive waste piles dwarf the surrounding topography.

Despite the waste, the slate industry grew rapidly after the Civil War and by 1876 American quarriers exported slate back to England. Few statistics exist prior to 1879, but between then and 1900 production tripled from 368,000 to 1.2 million squares of roofing shingles. One square equals a hundred square feet. Slate roof production remained above a million squares per year until World War I. Throughout the years of peak quarrying, Pennsylvania generally produced twice as much slate as the next most productive state, Vermont.

The vast majority of slate roofs are on the East Coast because the principal slate roofing quarries were nearby. Slate covers famous structures such as the original Smithsonian building and Vanderbilt's Biltmore House, and institutional buildings such as churches, schools, and government offices, but most slate went for commonplace structures such as barns, shacks, and homes. There's even a slate-roofed outhouse, built of wood with a Peach Bottom roof, in the old Welsh mining hamlet of Coulsontown, Pennsylvania.

In the American West, very few homes sport slate roofs. They primarily occur on institutional buildings. For example, the older buildings on the University of Washington campus, in Seattle, have slate roofs. All of the shingles came from the East Coast.

I didn't discover slate roofing until Marjorie and I moved to Boston. My first memorable encounter was at Harvard University. Ornate and somewhat gaudy, Harvard's Memorial Hall features a bold, striped pattern of green, red, and black slate shingles. Most are the traditional square-cut pattern, but about halfway up the four towers, builders placed rows of green diamond-shaped shingles. Because this shape was not as watertight, builders used diamond shingles on steep sections, such as on mansard roofs. Memorial Hall's builders also used hexagonal shingles, which work better on less-steep roofs. The building stands out for its beauty and garish colors and shapes, especially in contrast to the stark concrete slabs of the nearby science center.

One of the great pleasures of living in the Northeast is exploring small New England towns and seeing the colorful slate roofs. They come in shades of green, red, black, and purple with patterned and random blends of color. Some ambitious owners went to the trouble of laying the shingles so contrasting colors would shape out the numbers of the year the roof was installed. One of the oldest dates from 1851, still with original shingles. Not all buildings were in good shape; a run-down barn near a friend's house in Vermont looked as if the wood framing had pooped out and could barely support the weight of the slate. I suspect that in a few years the roof would be all that remained visible, leaving a disturbing scene like that of the Wicked Witch's hat after Dorothy tossed water on her. No matter the state of upkeep, slate roofs are part of what gives architecture in the East an enduring elegance.

Slate sold well until World War I because it was abundant, nonflammable, durable, and fashionable. Wood shingles were the main competition. They were cheaper than slate and generally available wherever trees grew. Wood had one disadvantage, its propensity to ignite. Neither slate nor wood, however, could compete with the rise of asphalt shingles, which consist of a mat of glass fibers coated in asphalt and covered in a protective layer of granules. By 1947 asphalt roofing outsold slate 437 to 1.

Asphalt cost less and was transported more easily, and asphalt roofing sellers marketed their product better. According to Jeff Levine, the asphalt men further benefited from the parochial nature of slate quarrymen, who failed to adapt to new technology, rejected the advice of geologists, and jealously guarded their trade's skills. Installing a slate roof also required skills far beyond the rip-it-up-slap-it-down techniques of an asphalt roofer.

On a typical house, a couple of guys can start in the morning, tear off an old asphalt shingle roof, and have a new one installed in a day or two. No slate roofer could or would attempt such a feat. They can put on one or two squares of slate per day, compared with a nail gun–toting asphalt roofer's daily output of twenty squares. Not that nail guns have increased speed—experienced asphalt roofers of old could install twenty squares per day with a hammer—but the mechanical nailers have made it possible for novices to crank out roof after roof.

"I prefer working with a new guy than working with an experienced composition roofer when installing a slate roof," says Ben Kantner, who has been installing slate roofs for over twenty years.[13] "The new guys are slower and more careful. It takes skill to set a nail right on a slate roof. The comp guys tweak and pop a lot of shingles because they bang in the nail." Composition or asphalt roofs require pounding because the nail secures the shingle in place. Slate shingles, on the other hand, hang on the nail and can float, or move slightly. To facilitate proper nailing, the nail holes are beveled. A correctly sunk nail nests in the bevel, neither sticking above the slate and preventing the next overlapping shingle from laying flat nor going too deep and cracking the shingle. Slaters also have to be able to hit a nail straight because they often use copper nails, which bend more readily than nails made of stainless steel.[14]

Slate roofing requires another precision skill, the ability to cut a shingle for spaces such as ridges, valleys, and chimneys. Slaters employ either a modified paper cutter, like those used by trimmers in slate mills, or a slater's stake and hammer. With the stake and hammer, they lay a shingle on the T-shaped stake and chop it with the thin edge of the hammer's shaft. Slaters have to be careful because the edges of slate are extremely sharp; Kantner needed seventeen stitches when a shingle sliced into his hand. In contrast, cutting an asphalt shingle requires no greater skill than the ability to operate a utility knife, part of the reason that a construction nincompoop like myself was able to reroof our house with asphalt shingles, although it took almost a month.

A slate roof costs significantly more than a comp roof. Installed asphalt shingles range from $50 to $150 per square versus $500 to $1200 per square for slate, with additional installation costs of $12 to $30 per square foot. Cost depends on color, with red the most expensive, and size, with larger shingles costing more. Slate shingles sell by size and range from six inches by ten inches to fourteen inches by twenty-four inches.

"People would rather put their money in the interior than on the exterior of their home," says Kantner. "From my point of view the exterior is more important. That's what protects the interior." But since Americans are mobile people we don't think in the long term and if a roof lasts only fifteen or twenty years, we don't worry because the leaks won't be our problem. We will have moved on. Contemporary consumers seem to care little about permanence. Slate and wood shingles were ideal products in the 1700s and 1800s, when people moved less. Having a good roof over your head was like having a good root cellar. You were planning for the future.

Using asphalt instead of slate reflects the widespread change in not using stone as much for building. Slate had been the most practical and popular roofing material, but the stone is so expensive to buy and install that only the wealthy can afford to use it now.

Because of the dearth of slate roofs in most areas outside of the East Coast, about the only place to encounter slate regularly is at a pool hall. If you could look under the green felt of a pool table, you would find slate. First used in 1826 by English billiard table maker John Thurston, slate has remained the unrivaled material of choice for pool tables. Thurston's original slate came from the Penryhn quarries in Wales. He chose slate as a replacement for oak—the material he had used for a billiard table sent to Napoleon in his exile on St. Helena—because it didn't warp, could be honed glassy smooth, and weighed enough to prevent the careless, or unscrupulous, from bumping the table and repositioning balls. Slate also was preferred because of its low cost.

Pool arrived in the United States in the late 1700s, but not until the mid-1800s did it lose its unsavory reputation.[15] By then, several companies had begun to fashion tables, including the J. M. Brunswick Company, eventually the largest billiard table maker in the world. Brunswick originally obtained their slate from quarries in Pennsylvania. No billiard table manufacturer, however, has used American slate for decades. It's too expensive. Everyone imports their slate from Brazil, Italy, and where else, China. It arrives precut and predrilled. Each table consists of three pieces of slate, cut from one larger slab. All manufacturers use multiple pieces because one big bed is too heavy and awkward to work with and too hard to keep level.

I bring up billiards in part because of its reliance on slate, but also because a pool table led to the creation of one of the few beautiful pieces of slate I have seen. It is a relief sculpture of four ravens by Seattle sculptor Tony Angell. He had wanted to work in slate for many years but didn't have the opportunity to do so until he pulled into a school parking lot one day and saw an abandoned billiard table. He salvaged the three pieces of the bed, taking home the one-inch-thick slabs that weighed six hundred pounds. (Robinson Jeffers also recycled billiard tables; slabs from Fort Ord form part of the path at Tor House.)

The ravens face to the right. Two have their heads raised slightly, perhaps alerted by the call from the two that have their bills open. Such a scene daily occurs in most urban parks. Exploiting the slate's fine grain, Angell carved exquisite details—shaggy throat feathers, reptilian feet, and fine feathers on their beaks. He also polished each bird so that they glisten like real ravens illuminated by the sun. Angell further highlighted the ravens by dulling the background with stippling and chipping along the cleavage. The patterns give the relief a surprising depth. His sculpture brings to mind Michelangelo's slaves; this is clearly stone.

The relief of the ravens reveals another unusual aspect of slate. It is the only building stone used primarily in two dimensions. Slate is basically a material of length and width. Consider its primary uses: roofing, blackboards, and billiard tables. Each use exploits how one can separate slate into thin sheets. You will not find slate foundations or walls, places that require length, width, and depth. Coquina, for example, worked as a building material primarily because it could be cut into massively thick blocks. Builders don't carve slate for three-dimensional embellishments, such as capitals or pediments, which makes it more remarkable that slate serves so many purposes even though we don't take advantage of all its dimensions.

The historic use of slate in schools and homes, and for recreational purposes, ensured that you would encounter slate every day, but even after death slate could still play a visible role. Boston's King's Chapel Burying Ground and most other old burial grounds in the East are great places to see slate.

King's Chapel contains several hundred slate tombstones in an area about as large as two basketball courts. The slate grave markers, now in several rows, were moved from a more haphazard setting into their present position by the superintendent of burials in the early 1800s. He thought it would make the grounds more attractive. He also placed headstones along the paths, giving you the ominous feeling that you are being herded unceremoniously across numerous graves.

Generally short and rectangular, the gray markers appear to have been relocated during a drunken binge. The superintendent got them into more or less straight lines but they lean left, right, forward, and back. A few have succumbed to gravity and tipped over completely.

Along one of these paths stands—or, more accurately, slumps—a slate grave marker with a three-lobed top. To reach it, walk through the burying ground's wrought-iron gate, turn right on the main path, swing left around an octagonal enclosure, and continue down the path next to the chapel to just beyond a tree with well-furrowed bark. The slumping stone rises on the left side of the path, next to a similarly shaped marker that has sunk half again as deep.

Originally, two slightly raised shoulders flanked the taller stone's tympanum, or middle lobe, which takes up the middle half of the grave marker. Gravestone researchers refer to this shape as a headboard. Prior to the superintendent's beautification project, another headboard-shaped but much smaller marker, called the footstone, stood about five or six feet away from the headstone. The deceased's coffin would have lain on its "bed of death" between the two markers and faced east, so the body could rise toward dawn at resurrection.

Carved into the tympanum is a winged skull, or death's head, with perfect teeth, as if death had seen an orthodontist. Atop the grinning skull flies a winged hourglass, about half the height of the skull. A rosette and garlands that resemble abstract owls run down the outside quarters of the panel below the lobed shoulders. Cut into the smooth center of the stone are the facts: Elizabeth Pain, wife of Samuel, died November 26, 1704, age near fifty-two. The words appear next to a heraldic shield, or escutcheon, bearing two lions, and several one-inch-wide lines, which link together in a resemblance to the letter *A*.

Elizabeth Pain's tombstone has a notorious reputation. In the final lines of his romance tale of morality in Puritan Boston, *The Scarlet Letter*, Nathaniel Hawthorne wrote: "In that burial ground beside which King's Chapel has since been built . . . [O]n this simple slab of slate—as the curious investigator may still discern, and perplex himself with the purport—there appeared the semblance of an engraved escutcheon. It bore a device, a herald's wording of which might serve for a motto and brief description of our now concluded legend; so sombre is it, and relieved only by one ever-glowing point of light gloomier than the shadow:—ON A FIELD, SABLE, THE LETTER A, GULES."

Do those linked lines on Pain's tombstone form the famed scarlet (*gules* means red) letter *A* on its sable background? Was Pain the model for

Elizabeth Pain's slate tombstone, King's Chapel Burying Ground, Boston.

Hawthorne's adulteress Hester Prynne? Did Pain's gravestone seed Hawthorne's imagination? Many people have raised these questions. The facts are few, the speculations many.

Hawthorne lived in Boston twice. The first time he lasted six months as editor of *American Magazine*. He returned almost three years later, in March 1839, and stayed until November 1840. Scholars know that during his time as editor, Hawthorne often visited the Boston Athenaeum, a famed library originally located next door to King's Chapel Burying Ground. A vigilant researcher and active explorer of Boston, he more than likely walked through the graveyard and saw Pain's gravestone. Adding a bit of spice to the story, Pain did go to trial, not for adultery, but for murdering her child. She was found not guilty, but still was whipped twenty times.

Many guidebooks and Web sites report that there is no doubt that either Pain or her gravestone inspired Hawthorne, but no one knows for sure. Although Hawthorne did base many characters in *The Scarlet Letter* on real people, no direct, unequivocal evidence links Pain and Prynne. Still, Pain's gravestone offers numerous reasons to visit it, for it exemplifies trends in stone, shape, symbolism, and language found throughout graveyards in Boston and beyond.

Slate was the stone of choice for gravestones for more than 150 years in eastern Massachusetts. The earliest carvers used wood, followed by local rock, often field boulders, before turning to the more abundant local slate. Well-known quarries opened in Cambridge, Slate Island in Boston Harbor, Harvard (twenty-five miles west of Boston), Charlestown, and Braintree (near the Granite Railway Quarry).[16] The latter quarry produced the infamous, twelve-inch-long trilobites that Percy Raymond described as bordering on the "domain of romance." None of the Massachusetts quarries are active at present and in a monumental 1914 United States Geological Survey study of slate, none are listed, but for two centuries they produced thousands of tombstones.

It is easy to see why early gravestone carvers chose slate. They could split the slate into slabs and could cut the stone with ease and with exquisite detail. Who knows if the carvers knew how well slate resisted erosion, but they probably would be surprised and pleased that you can still discern individual teeth on Elizabeth Pain's winged skull. Furthermore, you can make out the remaining parts of the whorl on the broken right shoulder, even though only the top layer of slate remains, and there is no doubt that lions populate the escutcheon.

The tripartite headboard shape of Pain's tombstone is a classic style of gravestone from Puritan New England, akin to what biologists call an indicator species. Whenever you see a headboard-shaped slate grave marker, you can almost guarantee that you are in a former stronghold of Puritans. In addition to the bed of death connection, a tripartite shape represented the arched doorway that the soul passed through on its way to eternity. Variations on the theme occurred, but for the most part the headboard remained popular to around 1800, when a neoclassical revival arrived and gravestones became taller, sleeker, and more three-dimensional.

A winged skull also appeared widely on gravestones during the Puritan era. Like the headboard shape, winged skulls told of death, arguably the central preoccupation of an average Puritan's life. They died young. They saw death regularly. They heard about death weekly from their preachers. Placing a winged skull prominently on grave markers reminded all who visited that their life would end soon, too.

And if Elizabeth Pain's visitors didn't get the point of the skull, they should at least have understood the winged hourglass. Time flies, your time is short. Subtlety was not a strong point of Puritanism.

Below the winged skull, Pain's gravestone reads "Here Lyes Ye Body of . . ." The words refer specifically to the mortal remains; yet another reminder that the body and not the soul was the focus for the Puritans. Although Pain's gravestone lacks an epitaph, many stones had ones that emphasized rot and decay, or as Benjamin Franklin wrote in 1723 in his mock epitaph, your body will be "food for worms."[17] No wonder modern people have such a gruesome image of the Puritans.

The imagery, slate, and wording of Pain's tomb were the common language of burial grounds throughout the 1700s, but with the decline of Puritanism and corresponding rise of more liberal religious viewpoints, change spread across the world of death. Less forebidding winged faces or cherubs replaced the ominous winged skulls, followed in the early 1800s by the rise of the urn-and-willow design, symbolic of more secular beliefs. In addition, "In Memory of" began to supersede "Here Lyes Ye Body of," as the focus of death turned more toward the mourner and away from those who died.

These religion-influenced changes coincided with overcrowding and poor burial practices at Boston's burying grounds. The latter problem led to grave robbers regularly poaching parts and bodies for medical studies and to diseases spreading with the malodorous miasmas wafting from the dead. New religious outlooks, a concern with sanitation and safety, and the rise of neoclassicism ultimately are made manifest in the development of a new type of burial ground, the cemetery.

Boston's elite, many of whom were helping pay for the erection of Bunker Hill Monument, established Mount Auburn Cemetery in 1831. Located four miles from Boston, Mount Auburn consisted of seventy-two acres of trees, dells, ponds, and wetlands. It was beautiful, free of bad air and grave robbers, and organized so that people could purchase lots, everything the Boston burial grounds were not. Reflecting the new religious views, Mount Auburn was a landscape where "death would seem disrobed of half its terrors," or so wrote an early visitor.[18]

Mount Auburn's founders took two additional tacks in their goal of exorcising the ghosts of Boston's old burial grounds. First, they banned perpendicular slab gravestones. Instead, they wanted three-dimensional monuments, such as sarcophagi, obelisks, mausoleums, and columns, and their allusions to antiquity. The founders also prohibited the use of slate; they considered it stiff, ungainly, and gloomy. And who wanted to be reminded of those dreary Puritans?

What did the founders replace slate with? Boston's Brahmins preferred marble, despite its reputation for poor durability in a cold climate. In contrast to slate and its connection to the Puritan's dour imagery of death, marble "suggested the purity of heavenbound souls and the assurance of salvation provided by liberal religion."[19] Unable to compete with purity and salvation, utilitarian slate's reign in New England graveyards came to an end.[20]

In February 2008 I returned to my old elementary school for the first time in over thirty years. Stevens was still an elegant Georgian-style building, but a new addition to the north marred the wonderful symmetry of old. Entering the front doors, I discovered that the wide wood steps that once led up to the second floor were gone. I stayed on the ground level and entered the school's office, what in my day had been the boiler room. I checked in, put on my Visitor button, and headed out in search of the blackboards of my youth.

As I got oriented, I found a stairway. It was much as I remembered, except now I noticed how thousands of little feet had worn a low spot in each stair about one foot from the banister. Someone, probably a well-meaning adult, had also placed metal nubs on the banister to prevent kids from sliding down. At the top of the stairway the floor of the hall was still wood and looked to be original, with some planks running twenty feet long or more. Dark wood still framed the doorways to each classroom, but the little cloakrooms where we hung our coats had been removed.

I stuck my head in one classroom. The blackboard was gone. I looked in another room. Also empty of slate. A teacher wandered by and I asked her about the blackboards. "They were all replaced when the school was remodeled in 1999," she said. "They were real slate and I have about twenty small pieces of it at home. I don't know what I am going to do with it."

The Seattle school district, like many across the country, has been phasing out blackboards for years. Dust is to blame. It's bad for kids because the fine particles exacerbate respiratory problems such as asthma and allergies. It's bad for computers because chalk dust gets into the keyboards and other sensitive parts. It's bad for fashion because chalk dust gets on clothes and in hair.

Every classroom at Stevens now has a whiteboard instead of a blackboard. Teachers use pens instead of chalk, which gives them the luxury of multiple colors, in contrast to earlier monochromatic eras of white chalk, or perhaps yellow. Pens provide better contrast and higher visibility and don't produce all of that nasty dust that coated hair and clothes. Nor can anyone annoy his or her classmates by running their fingernails down the blackboard.

Despite the numerous benefit of pens over chalk, whiteboards are far from ideal. Chalk dust can be an irritant to some people, but generally the large particles settle out of your nose before they have a chance to get into your lungs; whiteboard cleaners, though, produce toxic fumes. The cleaners come with warning labels. Another drawback to whiteboards is the residue of writing that doesn't wash off, no matter how much cleaner you use. And when someone uses a Sharpie pen by accident, you cannot erase it from a whiteboard. Furthermore, half the time you try to use dry-erase pens they don't work because they have run out of ink. Then what do you do? Toss them in the garbage so they can end up as landfill. No one ever has to doubt whether a piece of chalk is usable, and when it becomes too small to write with, you can just carry it outside and bury it. It will soon degrade and disappear without a trace.

As I wandered the hallways at Stevens, I couldn't help but wonder if today's students are missing out. Where once they could have continued the centuries-old tradition of employing fossil sea critters to write or draw on a metamorphosed slab of fine-grained marine sediment, now they write on petroleum-based, plastic whiteboards with an odoriferous, chemical-filled pen. In a society where our failed connection to nature surely has contributed to our failed understanding of human impact on the land, the loss of slate and chalk is one more example of how we are taking nature away from children and replacing it with something artificial. Perhaps this connection was the most utilitarian aspect of slate and now it, too, is lost.

# 10

# "Autumn 20,000 Years Ago"— Italian Travertine

*After they had noted what a profusion of resources has been begotten by Nature, and what abundant supplies for construction have been prepared by her, they nourished these with cultivation and increased them by means of skill and enhanced the elegance of their life with aesthetic delights.*
—Vitruvius, *Ten Books on Architecture*, book two

*He'll study weird stuff that grew in his sink last week, for instance, bird droppings, a bit of arterial plaque, or his wife's cataract. His instincts are amazing, though. No matter how oddball, the things Dr. Folk chooses to look at often end up teaching us something about rocks.*
—Dr. Kitty Milliken, University of Texas

GEOLOGISTS DO NOT normally receive standing ovations when they make presentations at meetings. Typically, attendees clap politely, ask a few questions, and then the next presenter walks to the podium and begins his or her talk. In October 1992, however, geologist Robert Folk received a standing ovation for his fifteen-minute presentation at the annual Geological Society of America meeting in Cincinnati. He had titled his talk "Bacteria and Nannobacteria Revealed in Hardgrounds, Calcite Cements, Native Sulfur, Sulfide Minerals, and (yes) Travertines."

Folk's talk focused on his work from calcite-rich rocks in Italy, as well as the Bahamas, Utah, and Florida. He began by describing how he had used hydrochloric acid to etch, or eat away, surface material to reveal undisturbed layers of calcite. With this novel etching technique, Folk reported that he had been able to exhume microscopic (requiring greater than 20,000× magnification) bacterial bodies from the rocks, primarily from a type of limestone called travertine, and that the microbes had "emerged from the calcite like cadavers on Judgment Day."

He called his microbes "nannobacteria," in reference to their nanometer-scale size. (Folk prefers his double-*n* construction, but most nongeologists prefer "nanobacteria" or the less controversial—read as no indication of life—"calcifying nanoparticles." Folk's use of "nannobacteria" still rankles many biologists.) No one had ever reported and shown photographs of such microscopic organisms. Prior to Folk's work, the smallest recorded bacteria stretched 200 nanometers in diameter, or .005 the diameter of the proverbial head of a pin. In contrast, some of Folk's nannobacteria were as small as 25 nanometers or .001 the volume of previously described bacteria, although many ranged up to 150 nanometers. The bacterial bodies resembled "ears of corn" and had "great fossilization potential," he concluded.

Chris Romanek was one of those who attended Folk's 1992 talk. Now a geologist at the University of Georgia, Romanek was then a postdoctoral fellow at NASA. "I had sort of wandered in, not knowing who was speaking," said Romanek. "I remember he was talking about using an SEM [scanning electron microscope] to see microorganisms in these travertines from Italy. I thought this is kind of interesting."[1] When Romanek finally learned who was speaking, he knew at once who Folk was: "I had used his textbook in college."

First published in 1974, *Petrology of Sedimentary Rocks* has long been a bible to geologists studying the rocks that make up 80 percent of Earth's crust, including limestone, sandstone, and coquina. Out of print for years, it is now available online from the University of Texas in Austin, where Folk taught geology from 1952 until his retirement in 1990. Folk wrote the book to supplement lectures and labs that he gave in Austin and based it on lectures he had attended at Pennsylvania State College, where he received his bachelor's, master's, and doctoral degrees. The book is so legendary that geologists have been known to come up to Folk at geology meetings and ask him to autograph their dog-eared copies.

"Folk's talk got me thinking about limestones and how they would be conducive for trapping and fossilizing microbes," said Romanek. Although his specialty was carbonate rocks such as limestone that contain the minerals calcite, dolomite, and aragonite, he had not thought about this connection between physical and biological processes in their formation. It was a connection that would have profound implications for Romanek's research.

Thirteen months later, Folk's talk came back to Romanek when his colleague at NASA, David Mittlefehldt, asked him to look at pictures of orange blobs on a meteorite from Mars. Mittlefehldt told Romanek that the microscopic, pumpkin-colored rosettes were carbonates. The photographs stunned Romanek, who had never seen carbonate minerals on a rock from outer space. He immediately asked Mittlefehldt if he could get a sample from the rock, which had been found in Antarctica in 1984. Romanek believed he had the tools, primarily through Folk's etching technique, to tease out the answer to the underlying question of how carbonates—minerals often associated with water and living organisms—developed on a potato-sized rock that had traveled millions of miles across space.

After receiving a pinhead-sized chip of the Martian meteorite, Romanek etched his sample with acid and began to probe the rock. Looking under the SEM, Romanek found a screen filled with bacteria. He couldn't believe what he was seeing, life forms on a rock from outer space. Unfortunately, he wasn't seeing microscopic ETs; instead, Romanek was looking at Texas bacteria. The microbes that polluted his sample came from unfiltered water he had used to dilute his etching acid. He tried again, this time with filtered water, and again saw features that looked like relicts of bacterial organisms through the SEM. The rod- and ball-shaped structures resembled the nannobacteria Folk had described from travertine at his 1992 Cincinnati talk.

Romanek knew that he had found something extraordinary within the meteorite's orange blobs. Since the rods and balls were discovered within carbonates—minerals known to form via biologically induced deposition—the structures could be organic remains. For the next two years, Romanek and his fellow researchers at NASA probed, lasered, and scoped the meteorite to determine whether Romanek's initial observations were correct. They also reviewed Robert Folk's extensive work on Italian travertine, in particular a paper he published in 1993 that showed two dozen photographs of nannobacteria.

On August 7, 1996, NASA astounded the world when they reported what Romanek and his colleagues had discovered in their Martian meteorite from Antarctica. The 4.5-billion-year-old rock, known as Alan Hills 84001 (ALH84001), contained evidence for life on Mars.

Chris Romanek's observations on nannobacteria-generated carbonates were central to NASA's argument. "If I had not seen Bob Folk's talk, I wouldn't have thought to do what I did," said Romanek. "His work definitely kick-started the whole NASA project." And Folk's work all started with travertine, a building stone quarried in Italy for more than two thousand years.

Roman architect and writer Marcus Vitruvius Pollio may never have given a fifteen-minute-long talk at an annual meeting, but like Robert Folk, Vitruvius was a close observer of Italy's sedimentary rocks. In his landmark *De Architectura*, or *Ten Books on Architecture*, he devoted his second book to building materials. "I thought that I should expand on their varieties and the criteria for their use, as well as what qualities they have in building, so that when this information is known, those who are planning to build will avoid mistakes and assemble supplies suitable for buildings," wrote Vitruvius sometime between 31 and 27 BCE.[2]

Vitruvius based his work on empirical investigations he made of the building stones quarried near Rome. In the category of soft and yielding rocks, he included the olive gray *Tufo Lionato* erupted from the Alban Hills southeast of Rome, and *Tufo Giallo della Via Tiberina* (yellow tuff from the Tiber Road), erupted from Monte Sabatini, north of Rome. Tuff is rock composed of volcanic glass fragments, crystals, and rock fragments deposited by pyroclastic flows produced by violent eruptions of magma. In Rome tuff forms resistant layers in the city's celebrated seven hills, one of which—Palatine—provided a source for building material in the fifth and sixth centuries BCE.

Conversely, Vitruvius's hard, enduring stones correspond to lava flows that crop out near Rome. Roman builders often used these lavas for paving stones. They still pave many streets in the Eternal City. Locals call the three-inch-square paving blocks *San Pietrini,* "little Saint Peters," playing on St. Peter's role as the rock of Christianity. Vitruvius also described rocks of moderate durability, including travertine.

Travertine is a sedimentary rock rich in calcite and the related mineral aragonite. Layered, dense, and well-cemented, travertine forms primarily in hot springs. The Romans used it as early as the second century BCE for voussoirs to strengthen the *Pons Mulvius*, which spanned the Tiber. They quarried most of their travertine in Tibur, now Tivoli, about twenty miles east of Rome. During Vitruvius's day, travertine was known as *Lapis Tiburtinus*, "stone of Tibur," later shortened to *tiburtino* and corrupted to *travertino*, travertine in English. Modern quarries at Tivoli supply most of the travertine used in the building trade; it is the most commonly used and commercially valuable building stone in the world.

Although the Roman tuffs had the virtue of being easily worked, Vitruvius wrote that "if they are put in open, uncovered places, then once they have been saturated with ice and frost they crumble apart and dissolve. Likewise, along the seashore they will wear away, eaten by the salt." In contrast, travertine "endure[s] every strain, whether it be stress or the injuries inflicted by harsh weather." To deal with the weaker tuffs, Vitruvius recommended builders quarry the stone in summer, let it sit in the open for two years, use only the best, and dump the crappy stone in with foundations. Builders could further protect tuff from water by covering blocks with plaster or by placing travertine on top of the blocks.

The Theater of Marcellus, built in 13 BCE, exemplifies the building techniques described by Vitruvius, according to Marie Jackson, a geologist who has written extensively on the technical expertise of Roman builders.[3] For the three-story structure, the Romans selected travertine for exterior arches and carved columns. The stone is scarred and cracked but still retains a sturdy grandeur. For the interior wall, the builders used brown tuff for the arch shafts, but in locations that had to withstand the greatest stresses, such as imposts and keystones, they chose travertine. To further protect the tuff, builders most likely applied thick stucco, which has long since eroded away. Less famous than the more centrally located and more imposing Colosseum, the Theater of Marcellus epitomizes a sense of the majesty of ancient Roman construction, the technical mastery of the engineers, and the beauty of its building stones.

Travertine building blocks, though, have a weakness. "They cannot be safeguarded against fire. As soon as they make contact with it, they crack apart and fall to pieces," wrote Vitruvius. (Like the Greek philosophers Aristotle and Plato, Vitruvius considered that matter consisted of unique combinations of the elements water, earth, air, and fire, which gave an object, such as a stone, unique properties. Too much air and fire, according to Vitruvius, made travertine susceptible to breaking at high temperatures.) Modern scientists point to the unequal amounts of extension and contraction along internal crystallographic axes in calcite for travertine's poor performance in fires. Jackson found, however, that tuff survives fire better than travertine because of its porous texture, which allows tuff to expand when heated with far less fracturing than travertine.

Fire often plagued ancient Rome. When the Gauls conquered it in 390 BCE, they burned Rome. Major and minor conflagrations hit once or twice a decade for the last two hundred years of the millennium. To combat fire at the Forum of Augustus, builders used an olive gray tuff called *Lapis Gabinus*, for a one-hundred-foot-high boundary wall. The wall still stands although Rome burned in 64 CE (Nero's great inferno), followed by five large fires over the next two centuries.

Recent geological work by Jackson and her colleagues has confirmed that Vitruvius made astute observations of the strengths and weaknesses of Rome's local building stone. She compared the durability of tuff and travertine in dry, humid (foggy), and wet (rainy) conditions, as well as the two stones' ability to withstand carrying a compressive load. The tuffs acted more like sponges and absorbed varying amounts of water, some to the point of crumbling, whereas travertine was more or less impermeable and retained its strength in wet conditions. Because of these qualities, travertine played a central role to Roman builders both as a structural reinforcement to tuff construction and as a decorative and protective facing on tuff walls.

Nowhere are travertine's attributes better seen than in arguably the most impressive building of ancient Rome, the Amphitheater of Flavium, or the Colosseum, as it came to be known in medieval times. Started in 70 CE by Vespasian and finished a decade later by his son Titus, the Colosseum contains 3.5 million cubic feet of travertine, or eighteen times the amount of Salem Limestone that covers the Empire State Building. Archeologist Janet DeLaine has estimated that in order for that much travertine to be transported the twenty miles by oxcart along the Via Tiburtina from Tivoli to Rome, one cart carrying half a ton of rock had to leave every four minutes. She concluded, "I leave it to the reader to work out the implications for keeping the roads clean."[4]

Despite offering such potentially pungent numbers, DeLaine suggested that most travertine reached Rome by the Anio River (now known as the Aniene). Barges would have been able to carry large quantities of stone down the calm, meandering river from the quarries at Tivoli without the need to acquire, feed, and take care of hundreds of oxen.

Few, if any, barges would be able to descend the Aniene now. For a recent event focusing on river pollution, an Italian environmental organization highlighted a green space on the Aniene about ten miles from Rome. The group had brought in a backhoe to remove trash, mostly plastic debris, but also refrigerators, wood paneling, and metal file cabinets, that covered the Aniene in a fifty-foot-long raft from bank to bank. Not that the modern Via Tiburtina is much more appealing; the road runs for miles past car dealerships, fast-food restaurants, and stolid apartment buildings.

Modern travelers can speed from the Tivoli quarries to the Colosseum via bus and subway, which drops you on a low rise with a fine view of the long and more intact north side of the amphitheater. From the vantage point on the hill, you can see how the massive stones play a central role in providing structural support. Three levels of travertine arches rise to support a final, unarched story of squared blocks of travertine. There are no false walls here; the building's exterior skeleton stands because of the interplay of stone and arch. You can also observe how traffic has blackened the stone with pollution, a problem found in travertine-clad structures around the world.

To reach the travertine, descend the hill and cross Via dei Fori Imperiali. Once across you can touch the stone. The walls feel pitted and eroded from centuries of weathering; the steps and walkways feel smooth from millions of feet. For nearly two thousand years, this awesome building has dominated and graced Rome.

A clockwise circumnavigation takes you around to the entrance and a quick security check. Travertine building blocks make up most of the outer corridors. In the open amphitheater, however, travertine gives way to brick and tuff, particularly in the walls of the substructure and the *cavea*, or seating area. The brick and tuff look refined and well chosen, but the travertine arches of the outer walls are the glory of the Colosseum. Masons cut each block on site so that keystones, voussoirs, and imposts fit together as an intricate puzzle, where each piece balances another. In many blocks, you can still see the holes where builders used forceps and levers to lift the blocks into position. (Other holes in blocks indicate where thieves excavated iron dowels and clamps that once held blocks together.)[5] On many arches the builders placed the blocks with their bedding planes aligned with the curve of the arch so that the beds look like rays shooting outward. Each arch is a small-scale illustration of the wonderful marriage of Roman geology, engineering, and art.

Not everyone has appreciated the Colosseum. Starting in the 1300s and continuing for nearly five hundred years, the amphitheater served

Arches at the Colosseum, Rome.

Rome as a quarry. One Giovanni Foglia of Como received permission in 1542 to remove 2,522 cartloads of travertine. Particularly egregious was the stealing of stone by the great families of Rome, who used the travertine for the Barbarini, Venezia, and Farnese palaces, each of which still stands. In 1540 Cardinal Farnese's uncle, Pope Paul III, gave his nephew permission to ransack the Colosseum for twelve hours. Farnese brought four thousand men to help him. Although no direct evidence exists, one who most likely benefited from the cardinal's travertine transgression was Michelangelo, who took over the design of the Farnese palace in 1546 and completed the building in travertine.

No other building may be better known in Rome for its travertine than the Colosseum, but you cannot travel more than a few minutes without stubbing your toe on the stone. You can walk up the travertine Spanish Steps, stand with hundreds of tourists at the travertine Trevi Fountain, wait to cross streets on travertine sidewalks and curbs, admire Bernini's oval, travertine Sant'Andrea al Quirinale, and drink water out of a splendid travertine fountain carved in the shape of stacked books. Travertine is the most common building stone in Rome. After two thousand years of use, travertine has more than lived up to Vitruvius's observations of its durability and suitability for those who want to avoid mistakes when building.

Robert Folk's fellow geologists describe him as controversial, eccentric, a publicity hound, and a hyperactive pseudo-elf, but they also agree he is a brilliant observer and a wonderful teacher, and that he draws original scientific connections. He also loves Italy—the food, the scenery, the arts, the music, the architecture. In 1979 he was able to combine his interests, energy, and passions in what would become a seminal study of travertine.

"Well, if you really want to know, I was working in Italy and looking for another excuse to stay," said Folk.[6] He was in Rome in the piazza of St. Peter's looking at Bernini's colonnade of Doric columns when he realized he couldn't place the rock. He knew it was travertine, but the stone didn't fit into the best-known classification system for carbonate rocks, which was odd because Folk developed that classification in a landmark 1959 paper. "This gave me an excuse to go look at travertine," he said.

To study the travertine, Folk teamed up with his former student Henry Chafetz, now a professor at the University of Houston. They traveled to Bagni di Tivoli (the Baths of Tivoli), the most famous travertine deposits in Italy and the source of Bernini's columns, the Colosseum, and most of Rome's travertine. The open pits are concentrated in a flat plain about three miles east of Tivoli and just north of the Aniene River. *Cavatori* during Roman times removed rock from an area next to the river still known as the Barco, a corruption of *barga,* Latin for barge, in reference to how travertine traveled to Rome.

Between fifty and sixty quarries operate in Tivoli at present. At the one owned by the Mariotti family, who have been quarrying the site for four generations, *cavatori* cut rock in a manner similar to what occurs in Indiana, using diamond wire saws, gargantuan front-end loaders and trucks, and gallons and gallons of water.

Travertine deposition at Tivoli occurred and still occurs in lakes, ponds, and swamps on a flat, volcanic plain pierced by thermal springs. Stone in the primary quarries is younger than two hundred thousand years with a concentration of deposition around eighty thousand years ago. Geothermal activity heats water deep underground and it rises to the surface via faults. Along the way the water passes through and dissolves

Mesozoic-age, calcite-rich limestone and transports the calcite in solution to the surface.

Similar to what happens when you open a soda bottle, the pressure drops when the heated water emerges at the surface, and the carbon dioxide escapes from the fluid into the air. The release in pressure increases the saturation state of the water with respect to the carbonate minerals aragonite and calcite, and after the water degasses, the minerals settle out of the water column. (To simplify the story, I will just refer to calcite.) Depending upon the environment, vent mouth or quiet pond, the calcite accumulates as a fine-grained mud, gloms onto other solid particles in the water, or fills voids.

Folk said of hot springs, "They can be very fickle. Some days they have an enormous flow when there's been a lot of rain and you get a very rapid precipitation rate of calcite. On drier days, you get a very slow precipitation rate, perhaps on the order of a tenth of a millimeter a day." He has recorded calcite deposition of one-sixth inch per day, or a million times faster than calcite accumulates in the ocean. Carbonate accumulation can occur so rapidly in travertine that Folk once described the ghastly scene of algae and mosses "being calcified while still alive." I don't recommend letting children see this shocking geologic process.

Hydrogen sulfide also escapes into the air above the hot springs, and in such great quantities that Folk and Chafetz saw many dead birds littering the pools at Tivoli. Chafetz remembers hearing that three boys died breathing the toxic fumes, a year or so after the geologists finished their study. The fumes smell like rotten eggs and are a classic sign of hot springs, such as at Yellowstone National Park, where travertine forms the falls at Mammoth Hot Springs.

Geologists refer to this degassing of carbon dioxide and precipitation of carbonates as an inorganic, or abiotic, process. For the last hundred-plus years, geologists accepted that this abiotic process fully explained how most travertine formed. But then in 1984 Folk and Chafetz published their study of Tivoli. Titled *Travertines: Depositional Morphology and the Bacterially Constructed Constituents*, it proposed the radical idea that bacteria bore primary responsibility for travertine deposition.[7]

They were not the first to recognize the connection between bacteria and calcite deposition. As far back as 1914, British marine biologist George Harold Drew proposed that bacteria helped generate calcite in the Bahamas, but not until Folk and Chafetz did anyone present such a thorough study linking significant carbonate accumulations, identifiable structures, and bacteria. "We made this dumb luck discovery. It was exciting," said Folk. "We didn't go there with any idea of finding bacteria at all. A lot of scientific studies are just plain dumb luck." Luck or not, the paper has become the most cited scientific study ever written about travertine.

Bacteria influence travertine formation by altering the microenvironment around themselves. First, they excrete carbon dioxide during photosynthesis, which provides carbon and oxygen, the main building blocks of calcite. Second, bacteria can process inorganic nitrogen and generate ammonia gas, which raises the water pH, makes the water supersaturated in calcite, and generates calcite deposition. The microbes must achieve a balancing act because if they induce mineral precipitation too quickly, they get entombed, "to the dismay of the bacteria," as a sentimental pair of later geologists wrote.

Bacterial-stimulated crystallization most often occurs in nonflowing water, where it produces branching, shrublike growths. The wee forests often form in ponds and can extend laterally for tens of yards with shrubs growing an inch high, although they can also skyrocket to three inches in the right conditions. Some forests even tilt in the direction of water currents. To extend the forest analogy, inorganic calcite precipitates on and around the shrubs, eventually accumulating deeply enough to bury the shrubs and to provide a substrate for the next layer of bacterially generated shrubs to develop. Chafetz and Folk hypothesized that each shrub layer, or bed, represents one year of deposition with maximum shrub growth in summer and maximum "snowfall" of calcite in winter.

By cutting the stone perpendicular to the bedding, quarrymen exploited the dense, well-cemented shrub-and-snow texture. In ancient times, builders took advantage of travertine's low compressibility by placing blocks with the beds horizontal. For modern travertine used as cladding and not for structural purposes, architects disregard geology and structural integrity and attach the thin sheets to walls any way that looks good, with the result that bedding planes may run vertically and not horizontally.

Architects don't appear to share geologists' affinity for travertine's characteristic holey texture, though the voids are a good place to see where calcite crystals have grown. When I lead geology-oriented walking tours in downtown Seattle, I encourage people to use a hand lens to explore the pockets and look for six-sided calcite crystals. (I have also been known to sprinkle balsamic vinegar on the crystals. Geologists more typically use dilute hydrochloric acid to create a reaction with the calcite, which bubbles off carbon dioxide, but I like balsamic vinegar; it's always available and has multiple uses.)

Builders don't like voids because they provide a good spot for pollution-related particulates to collect. The dirt makes many buildings with exterior travertine look blotchy, as if they had the bubonic plague. To counter dust and soot accumulation, builders often fill in the holes in travertine with grout, which also prevents damage from water pooling, freezing, and expansion.[8]

Reacting to Chafetz and Folk's radical shift in explaining how travertine formed, geologists from around the world tried to verify the Texas geologists' conclusion. Within a few years, bacteria had been reported from travertine in Germany, Idaho, Yellowstone National Park, and Morocco, as well as at other travertine deposits in Italy. All occurred in warm, chemically harsh waters. However, Folk and Chafetz had one significant problem with defending the conclusions in their 1984 paper; they didn't see any solid bacterial bodies in Italy, only the voids left behind by decayed bacteria. What was missing was good fossil evidence.

Folk returned to Italy in June 1988 to look for more bacteria in travertine. This time he traveled north of Rome to Viterbo and the Bagnaccio and Bulicame hot springs, both of which have been well known for their therapeutic waters since Etruscan times.[9] "Bob wanted to pursue what he called the 'lesser rocks,' those with complex histories that formed under enigmatic circumstances. They also happened to be the prettiest," says his former student Paula Noble, now a paleontologist at the University of Nevada, Reno.[10]

At Viterbo, Noble and Folk collected mud and samples of recently formed travertine. They also dropped coins and pipes in the water to test how quickly travertine formed. On the pipes, travertine precipitated faster in areas hit by sunlight, which were richer in bacteria. No travertine accumulated on the copper coins because bacteria don't like copper. "He was so excited to see these examples of how bacteria affected travertine," says Noble.

When he and Noble returned to Austin, Folk began to examine his samples with an aged electron microscope. At 5,000× magnification, the most powerful the old scope could manage, he saw beautiful calcite and aragonite crystals. Folk described them as "resembling an ear of corn with a tattered shuck enclosing it" and "covered with a forest of spikes like a fakir's bed." The scope also revealed calcified bacteria. He now had the evidence that had eluded him for many years. Better material was yet to come.

In early 1989, the University of Texas acquired a new scanning electron microscope. The SEM's 100,000× magnification revealed tiny spheres and bumps studding the calcite. Folk initially ignored them, as had other SEM researchers, thinking they were sampling artifacts or lab contamination. Not all of the samples, however, exhibited the microballs. In some samples the spheres clumped together. He continued to look at the odd balls but also retreated to the library, where he found descriptions of similar-looking objects that microbiologists called ultramicrobacteria, or dwarf bacteria. The microbiologists hypothesized that the dwarf bacteria had shrunk in response to toxic conditions, incompatible temperatures, altered pH, or nutrient stress. Upon a return to normal conditions they would revert to their normal size.

Folk questioned the biologists' hypothesis particularly after seeing an SEM photo from Bulicame in April 1990. The eureka photo is mostly gray with a white plane, the top part of the crystal, running from the upper left corner to the bottom right corner. Below the plane is a light gray surface dotted with many spheres and one white, sausage-shaped body. On the left side of the plane rests an ovoid body, which looks large enough to encompass most of the other spheres. The ovoid is about twice as tall as the most unusual shape in the photo—a thin, saguaro-cactus-like body that rises from the center of the plane and which Folk described as a chain of eight nannobacteria. The entire shot shows an area 3.5 microns wide by 2.5 microns high, less than half the size of a red blood cell.

To Folk, these odd little balls and chains closely resembled bacteria shapes, and didn't look like any minerals he had seen in his decades of peering through microscopes. Unlike minerals, which tend to grow together as one mineral and not remain as single entities, they also clustered in swarms like bacteria in a "feeding frenzy." The more Folk peered at these unusual objects, the more he became convinced he was seeing life and neither artifacts nor minerals. He published his results in 1993 in another landmark paper, *SEM Imaging of Bacteria and Nannobacteria in Carbonate Sediments and Rocks*.[11] This was the well-illustrated paper that Romanek and crew poured over in trying to understand Martian meteorite ALH84001.

Once he found nannobacteria in his travertine, Folk began to find them everywhere. They were in cold freshwater spring deposits, deep underground in caves, in shallow saltwater, and in two-billion-year-old dolomite. Nannobacteria-generated calcite clogged a pipe from a hot water heater and scummed a birdbath. Nannobacteria balls covered corroded iron, copper, aluminum, and lead. Noncarbonate rocks, such as opal, chalcedony, and chert, also showed a nannobacterial origin. Unknown to Folk, others had begun to see nannobacteria, too.

Folk was working in his lab on August 7, 1996, when a colleague rushed in and said "Have you heard the NASA TV conference? They found nannobacteria in the Martian meteorite." What Folk's colleague didn't tell him was that in describing their nannobacteria, the NASA team had used a slide showing the nannobacteria Folk had described from travertine at Bulicame. "Yes, I was excited, but when their results first came out in the press as photos, my colleague Leo Lynch and I both thought the same thing. Gee, their pics look like excess gold coating artifacts," said Folk. "Later when they used a thirty-second gold coat, we were convinced that they did indeed have nannobugs." When Folk was able to look at ALH84001 under a SEM he found it permeated with nannobacteria.

Not everyone agrees that the Martian meteorite teems with supermicroscopic life. One scientist I corresponded with e-mailed me: "The existence of nannobacteria in extraterrestrial rocks (or any other life form, for that matter) is completely UNPROVEN, in my opinion (an opinion that I think you'll find to be almost universally shared)." The skeptics' main concern is that classic line "size matters."

Nannobacteria are too small to contain all of the nucleic acids and ribosomes necessary for life. This was the conclusion reached by a distinguished panel of scientists in a National Academy of Sciences workshop in October 1998. Inspired by NASA's report of Martian life, the panel sought to establish the smallest size of a free-living organism. The consensus was that between two hundred and three hundred nanometers "constitutes a reasonable lower size limit for life as we know it." Folk's smallest nannobacteria are only fifty nanometers wide.

Despite the establishment of this Maginot Line on the minimum size of life, researchers continue to report that they have found and cultured, or reproduced, nannobacteria-like bodies. The hottest field at present is in medicine, where researchers have reported nanoparticles in human, rabbit, and bovine blood. As occurs in travertine, the nanoparticles appear to generate calcite that contributes to health problems as diverse as kidney stones, malignant tumors, Alzheimer's disease, and heart disease.

"At this point we cannot be sure if the nanoparticles are living or not. Part of the problem is that we haven't perfected how to investigate them," said Virginia Miller, a professor of physiology at the Mayo Clinic.[12] In 2004 she was lead author on a paper that examined the role of nanoparticles and atherosclerosis, and in 2006 she co-organized a conference on pathological calcification, which brought together experts in biology, geology, and medicine.

Although Miller still wavers on whether nanoparticles are living organisms or not, she, like everyone else I talked with, credited Folk with stimulating her research. "I think we are at the forefront of something exciting. I will accept whatever we find but either way Dr. Folk's work has really given us a new way to think about the disease process, at least in regard to kidney stones and arterial calcification."

Miller's comments get to the heart of Robert Folk's research with travertine, bacteria, and nannobacteria. Ultimately, it isn't critical whether the nanometer-sized particles he sees are living or not. What is important is the scientific process. He saw something, travertine in Bernini's columns at St. Peters, that intrigued him. He studied it and published his results, which raised awareness in others and more questions in him. He went back out in the field, collected more samples, and took advantage of technological advances to ferret out additional answers. He also was persistent, spending hours and hours learning to use the new technology. Again he reported what he had found, and he continued to try to better understand and learn about what he had discovered.

No place better exemplifies the use of travertine as a building stone in the United States than the Getty Center in Los Angeles. Designed by Richard Meier and built between 1984 and 1997, the Getty incorporates 290,000 travertine panels to cover the multibuilding complex of museum galleries, restaurant and café, auditorium, research institute, and conservation institute. All of the stone came from the deposits at Bagni di Tivoli.[13] The roughly one million square feet of travertine ranges from 7½-by-7½-inch paving blocks to fifty-by-ninety-inch picture, or feature, stones, as architect Richard Meier termed the largest panels. He scattered them throughout the complex, often at eye level, in order to "provide a heightened awareness of the material."[14]

The Getty sits on a high ridge just west of Interstate 405 as it heads north from Westwood and up to Sepulveda Pass. Parking is underground in the multilevel garage. A travertine-floored elevator carries you up to the tram station and platforms of travertine. Since visitors cannot drive up to the Getty Center, most everyone parks in the garage and takes the 4½-minute ride on an electric tram up the hill.

Meier has described the tram ride as elevating visitors "out of their day-to-day experience" and "at the same time, they'll have a powerful sense of being in the center of this great city."[15] Perhaps he meant that being stuck in a vehicle with a top speed of ten miles per hour would give you that true, L.A.-traffic-jam feeling.

The tram rolls quietly for three-quarters of a mile through the chaparral vegetation of the Santa Monica foothills to the summit, 881 feet above the Pacific Ocean. Detramming occurs in the arrival plaza, a flat, open square with three pines and a large sculpture, titled "That Profile." Map-wielding docents meet the trams. They may also offer umbrellas, not for rain, but for sun. In the summer, umbrellas must be essential, as the sun blasts and bakes the travertine.

The plaza is the first introduction to travertine in all its splendor. The walls are unlike any other travertine building stones in the country. Most of the thirty-by-thirty-inch blocks have a rough surface and look like sepia-tone aerial photos of the desert southwest with miniature mesas,

The Getty Center, Los Angeles.

plateaus, valleys, and ridges. Although most of the blocks are the color of lightly toasted sourdough bread, some have highlights ranging from brownish orange to black, which enhance the three-dimensionality of the panels. The color results from oxidized iron.

Because Meier associated the normal cut of travertine—which sliced the stone perpendicular to the bedding—with building lobbies, he wanted a different look for his stone. He achieved it by cutting the Getty's travertine along the bedding planes, akin to opening a deck of cards and seeing the faces. "We wanted the stone to look like stone when viewed from the freeway," said Michael Palladino, who ran Meier's Los Angeles office during the construction of the Getty.[16] "Up close typical travertine looks nice, but from a distance it looks like wallpaper."

"The process was fairly simple. We would cut thirty-inch by thirty-inch by eight-foot-long sausages and shove them through the guillotine. The blade would fall and break the stone along its weakest point. We never knew exactly where it would split," said Palladino. The initial piece would be eight to ten inches thick, too thick to hang as veneer, so it would be cut down to thinner slices forming two façade stones and two to four pavers. "We were pretty proud of ourselves for figuring out how to cut the stone into so many pieces, but we were still way over budget," he said. Despite such cost-cutting measures the total cost of the Getty complex ended up being $1 billion.

Meier didn't originally plan on using travertine. He is best known for his metal panels, always white, which he has described as the "most wonderful color because within it you can see all the colors of the rainbow."[17] The Getty's neighbors, including the powerful Brentwood Homeowners Association, however, didn't buy such a flowery description of white. For six years, they wrestled with the Getty over the project. In the end, the two sides agreed to a building permit with 107 conditions, one of which banned Meier from working in white. Other requirements mandated a sixty-five-foot height restriction for buildings and that dirt could neither leave nor get shipped to the site. Meier described the neighbors' attitude as "we don't want to see you, we don't want to hear you, and we don't want to smell you."[18]

"The landscape also dictated that we use stone," said Palladino. "It is raw and rugged. We could either chop the hills and drop them flat or leave the topography and use this architectural element to connect the buildings to the land." If they had used metal, he added, it would have given the buildings a more pristine look and created an object quality; the buildings would appear to have settled into the landscape. "Stone, in contrast, would give a sense of the buildings rising from the land," he said. "We also wanted to give the Getty a sense of history and stability. We wanted to ground the institution, even though it was a young organization."

Palladino's observations are a bit ironic. In order to build the museum, the Getty shipped a hundred boatloads of travertine across the Atlantic Ocean, through the Panama Canal, and up the Pacific Ocean to Los Angeles. How strong is the connection to landscape, considering that the central grounding force, the travertine, had to travel halfway around the world to connect the museum's structures to place? I don't mean to single out the Getty for what they did; builders and architects have been shipping stone willy-nilly around the world for centuries. Such trade will not stop, and even seems to be increasing with new stones from exotic locales appearing every day, but transporting building stones has an environmental impact that builders and architects should consider.

To start the process of finding the right stone, Palladino sent notes out to stone suppliers asking them to send panels to the Meier office near UCLA. "This generated a whole lot of interest," he said. They literally had to rent a warehouse to store the hundreds and hundreds of samples. He and Meier eliminated 75 percent of the stone based on color, realizing after a more careful study and exploration of the site that they wanted a stone similar to the natural buff-colored rock on the building site.

Several stones stood out but were eliminated because they were too expensive or the quarry couldn't provide the quantity or provide it fast enough. "One quarry was excavating by hand. They couldn't have gotten enough rock to us in a hundred years," said Palladino.

They finally chose the travertine, in part because of the persuasiveness of Carlo Mariotti, the late father of the present owners of the Tivoli quarry. On a visit to Rome, Palladino and Meier had noticed rough-cut, nonbedded travertine used as decorative elements in buildings. They realized that this was the texture they were looking for, but no one had ever used this cut of travertine for a façade, particularly for a large project like the Getty. Mariotti convinced them he could cut the stone to show the rough surfaces. It took a year to develop the giant cleaver, dubbed Big Bertha, to cut the vertical slabs.

Carlo Mariotti was also the man responsible for supplying stone for many of the travertine buildings in the United States.[19] In 1958 he provided the stone for Ludwig Mies van der Rohe's Seagram Building. Although Mies used travertine only in the plaza and entryway, it provided a classic contrast to the modernism of the glass tower. More fame for travertine came between 1962 and 1968, when architects such as Philip Johnson and Eero Saarinen clad Lincoln Center for Performing Arts in travertine from Mariotti's quarries.

Most major cities in the United States contain at least one building dressed in travertine. Architects often use travertine as van der Rohe did, in foyers, as flooring and accents, the Sears Tower in Chicago and the former World Trade Towers being good examples. (The Towers' architect, Minoru Yamasaki, who coincidentally went to the same high school in Seattle as I did, had a close friendship with Carlos Mariotti and used his travertine in buildings in Japan, Boston, and Seattle.) The first travertine most people encounter in Los Angeles is on the walls and floors of LAX airport.

The Mariotti quarry and factory still use Big Bertha and its accompanying machinery. American architects, however, do not employ the cleft stone. "Big-name American architects won't work with it because they consider it to be 'Meier's stone.' Architects from other countries, however, don't have the ego problem and we sell a lot of it to builders in the Middle East," said Fabrizio Mariotti, who along with his brother Primo now runs the family business.[20]

As you wander the Getty and walk up to any wall of travertine, you will probably find fossil leaves in the cleft stone. They look like poplar, ivy, and dogwood and are so detailed and well preserved you can pick out individual veins and stems and tell if the leaf was upside down or right side up when it fossilized. A few even appear to have been torn or partially eaten, perhaps by leaf miner insects, prior to fossilization. The leaves look as if you can simply peel them off the wall.

The fossiliferous panels are ubiquitous: Meier thoughtfully placed many of the best in popular locations or in places such as along stairs where you have to move slowly. Some slabs have only a few leaves, but many have leaves piled atop each other. On the leaf-rich panels, the leaves face in all directions, just what to expect of leaves that fell in a shallow lake and collected together in the ooze on the bottom. Fabrizio referred to this phenomenon as seeing "autumn 20,000 years ago."[21] The autumnal feeling is enhanced where beautiful bronze and brown California live oak leaves have dropped next to the fossil leaves.

Not all of the travertine is sourdough colored. In the main entry to

Close-up of leaf fossil at the Getty Center, Los Angeles.

the museum buildings, Meier chose a chocolaty travertine to provide a contrast between the exterior and interior. A couple of panels, particularly ones near the bathrooms, feature an unusual set of fossils. One looks like a mini–Milky Way Galaxy, another like a tiny spaceship, complete with thrusters, and a third resembles a snail. Less than one-half inch wide, the white to tan fossils had been snails, most likely land-based species that had fallen into the water.

The most amazing panel at the Getty and the coolest part of the entire complex is one of Meier's feature stones. It is on the backside of a wall in the arrival plaza, directly opposite the tram station. The ninety-by-fifty-inch panel juts out ten inches from the surrounding panels and is the most stunning panel of building stone I have seen anywhere.

The giant slab of travertine looks like a jumbled mix of straws, melted wax paper, and spaghetti except that the straws are fossilized reeds, the waxy sheets are fossilized rafts of bacteria, and the spaghetti strands are fossilized mosses and algae. White to translucent calcite has filled in some reed tubes either completely or enough to leave only a small hole in the center. Most are hollow, though, and wide enough to stick a pencil into. The fibrous moss and algal strands cross and crisscross like a spiderweb.

As with the leaf panels, there is a distinct feeling of water; you can envision a quiet swamp with reeds pushing up through a surface coating of bacteria and moss. Some of the reeds stand upright. Others tilt and overlap. Perhaps birds or insects alit on the hollow green stems, seeking a meal of the invertebrates that lived on the mosses. (Coincidentally, on either side of the wetland panel are two much smaller panels, each with a fossil of a ten-inch-long feather.) You can almost hear the calling of frogs and smell the pungent decay of vegetation. The large brown panel is exquisite, life in a wetland caught in the act of fossilization.

Another texture at the Getty seems almost too ephemeral to have become rock. The surface of many blocks is honeycombed with one-eighth-inch-wide circles, but bees didn't produce the pockets. Instead the texture resulted from solidified gas bubbles, basically solid foam. As the bubbles floated through the water, calcite coated the surface and froze the diaphanous effervescence as a permanent feature. Because of the way the guillotine split the travertine, it chopped open the gas bubbles.

Calcite can also accumulate on the surface of ponds, building up playing-card-thin gossamer rafts. As happens with the bubbles, rafts can be blown across a pond and collect in thick layers, which in stone look like strewn sheets of phyllo dough. The raft-rich panels are widespread at the Getty and many also feature fossilized stems of green algae that resemble very thin, salt-coated pretzel sticks.

Folk and Chafetz's shrub forests are visible at the Getty, too. The best way to see them is to take a seat on one of the numerous travertine benches scattered throughout the complex. Leftovers from past centuries of cutting at Tivoli, the rough blocks enthralled Richard Meier when he visited the quarries; he thought they would contrast well with the more formal, cut travertine. If you look at the sides of the benches, which were sliced in the traditional perpendicular-to-bedding manner, you can find many layers rich in year after year of shrubs. You can also see the shrubs in plane or top-down view on the walls: Look for broccoli- or cauliflower-type patterns.

The seething environment at the mouth of the spring, as opposed to the quiet of shrub-generating pools, spawned egg- or pea-shaped balls, known as ooids or pisoids. They form as calcite gloms together in bands like a tightly packed snowball. Because the constant agitation knocks the calcite masses together, corners get removed, leaving behind the characteristic round balls. At the Getty they look like small mushroom caps.

This diverse array of textures exemplifies an important geologic theme—depositional environments are always far more complex than how most geologists describe them. For example, I can make the simple observation that the Tivoli travertine formed in lakes and ponds, but in reality those bodies of water have an incredible diversity. Water temperature, depth, clarity, chemistry, and gas content all vary, as do daily and seasonal conditions, surrounding vegetation, and long-term climate changes.

Out of this heterogeneous landscape came fossil leaves, honeycombs of gas bubbles, and a petrified wetland. No two panels are alike because no two parts of nature are alike. Rock excels as a building material because it bestows on buildings a complex beauty not found in materials such as glass, steel, concrete, or titanium. Each of these materials serves a specific purpose and can be beautiful in its own manner, but none prompts you to linger and look in detail. One metal Meier panel is just like the next metal Meier panel.

Stone bewitches because it is alive—a living, breathing material that changes gracefully over time. The softer Salem Limestone erodes around its harder fossils, creating a display case for the holy trinity of the Mississippian. Lichen and mosses colonize brownstone and contrast with another late comer, a blue black patina of varnish. Coquina weathers to an ashy gray and acquires hanging gardens of grasses and flowers. And the Getty travertine has already changed, losing some of its beige and becoming whiter. None of the human-made materials has a vitality like stone.

People further relate to rock as a building material because they intuitively sense the link between stone and the earth around them. Even if they can't tell the difference between granite and marble, they know that building stone has a history and a story. No manufactured building material can provide the deep connection to place that stone does.

Meier succeeded in his goal of connecting the Getty complex to the landscape, principally because of the rough-cut travertine. Similar to what happened at Robinson Jeffers's wonderful Tor House and Hawk Tower, the stone gives the buildings an organic feel. Meandering through the grounds feels like walking on a rocky hillside, seeking out beautiful fossils, attractive textures, and strange shapes. A friend who works at the Getty says that whenever he has rock-climbing friends visit him they always want to put up climbing routes on the walls. (It's a good thing I didn't have my rock hammer.)

You can, however, spend time rapping the panels with your knuckles. Los Angeles' seismic history led to a decision to bolt each panel about three-eighths of an inch away from each surrounding panel and from its concrete backing. This allows the panels to move during an earthquake. It also met Meier's contradictory goal of having stone that looks weighty and thick and is non-load-bearing.[22] In addition, hanging the panels resulted in an unplanned aspect of the Getty—music. Each panel produces a tone when you rap it with your knuckles, and because the panels vary in thickness, each tone is distinct. No other building produces such music.

Animals have also taken to the travertine. None has gone as far as the pigeons at the Castillo de San Marcos that started to eat the stone, but brown and blue speckled lizards regularly cruise the walls. In several areas, where the travertine touches the ground, holes contain the flotsam of insects and spiders.

Travertine helps make the Getty complex more accessible. The size and layout of the buildings are daunting and confusing but the travertine brings the focus back to a human scale. The stone engages visitors and encourages them to slow down and look more carefully. And even to touch. On one of the docent-led tours, our guide showed us a panel with fossil leaves. They were dark brown from all the fingers that had stroked them.

The Getty and the Colosseum share many similarities. Both were built on a scale to impress the visitor and dominate the site. Both required years of labor and hundreds of workers, although the Romans finished their project three years faster. Both now attract tourists by the busload, except at the Getty the tourists get a tram ride, too. Both are based on geometric shapes, the Getty on the square and the Colosseum on the arch, each repeated over and over again, giving the buildings their sight lines, their gravitas, their signature shapes.

Ultimately, though, travertine is what unites the Getty and the Colosseum. Both needed the stone in order to succeed as buildings. The Getty really didn't *need* it: Meier could have found another stone that would have allowed him to connect to place, but he needed the travertine aesthetically. The stone's textures, fossils, and colors give his buildings life, and travertine's Roman allusions connect the Getty to antiquity.

The Colosseum had to have travertine, too. No other contemporary material would have worked as well structurally. Tuff was too weak to support great loads and lava too hard to cut into precise arches, columns, and capitals. The Romans had the ability to import granite and marble, but not on the scale or at the cost required for the Colosseum. They also had concrete and did use it for vaults, but it had a tendency to creep or deform under great loads. And the nascent brick industry was unable to provide enough bricks for the building.

The Colosseum, the oldest building I visited, and the Getty, the newest, illustrate thematic end points central to the story of building stone. Builders started by exploiting their local geology and working with stone easy to carve and cut. They also considered whether and how the stone could provide structural support, especially with ambitious builders and architects, who wanted to build bigger and bigger. As time progressed, however, transportation replaced geology as a central driving force. Building decisions revolved around what a builder could get shipped to a building site. Good rail or boat service usually dictated this round of decision making. Finally down the road, money replaced transportation as the central part of the equation. The decision now revolves around how much money a builder is willing to shell out.

These themes of geology, transportation, money, and fashion play out repeatedly across time and geography. They reveal the timeless power of building stone and how we have used and continue to use it to convey sentiments as pedestrian or as grand as we desire them to be. For as long as humans have sought shelter, we have used stone. It is as elemental to our lives as water and fire and allows us to mark our place in the world.

From ancient Rome to Los Angeles to Mars, travertine exemplifies the complex stories of building stone. It is a symbol. It is history. It is science. It is a story in stone that one can see every day if we take the time to look, to ask questions, to wonder about the world around us.

# Acknowledgments

Like a conglomeratic rock, this book was assembled from many sources. One of the great pleasures in writing *Stories in Stone* was my interaction with geologists, historians, preservationists, and quarry owners and workers. They shared their passion, answered my endless questions, took me out in the field, and gave me samples. I will try to list all who helped but know I will inadvertently omit a few. I apologize for doing so. And, of course, any interpretations, errors, and opinions about their data are mine.

For the chapter on brownstone, my field time in quarry, museum, and the streets of Brooklyn with Alex Barrett, Alison Guinness, Mike Meehan, and Steve Sauter was essential, fun, and eye opening. The chapter on the Quincy Granite was the first one I wrote. The support of Jim Skehan, Tom Mahlstedt, Dick Bailey, Vic Campbell, and Richard Naylor set the tone for the rest of the book; they were generous with their time and patient in responding to my questions. Aaron Yoshinobu's fascination and passion for Robinson Jeffers, Tor House, and Hawk Tower were contagious and motivating. Dan Rea, Sarah Dodd, and Mark Gross of Cold Spring Granite Company went out of their way to help me see the Morton Gneiss and the Cold Spring mill. I also had helpful and numerous conversations on those ancient Minnesota rocks with Pat Bickford and Terry Boerboom: They helped clarify the science of this most challenging of stone.

When I began work on the coquina chapter, I received a wonderful e-mail from Leslee Keys. She made me feel at home, far away from mine, especially with that ice-cold gin and tonic. Joe Brehm's insightful tour of the Castillo de San Marcos made the fort and its history come to life. I also enjoyed a fascinating discussion and tour of the Florida Natural History Museum with Roger Portell. It is one of the best natural history museums I have seen.

As a native of Kentucky, I had some hesitation in heading north into rival Hoosier territory, but I need not have. Jim Owens made sure I met the right people and saw the quarries and mills. Will Bybee and Andy Chaney graciously opened up their quarry and mill for me, and Todd Thompson and Brian Keith provided the details I needed on the geology of the Salem Limestone. Of all the people I met while working on this book, few gave me as much help as Ruby Wilde and Carolyn Peyton; they tracked down obscure documents, regaled me with stories of early Lamar, and, finally, when I got to town, drove me, fed me, and introduced me to everyone I needed to meet, especially Greg and Val Emick, without whom I would never have seen petrified wood in the field, and Dorothy Smith, who provided such a memorable connection to the little gas station of her youth.

My dear friends John Horning and Terry Flanagan were great travel companions in Italy. From strolling to drinking wine to seeing endless rocks, they were enthusiastic, supportive, and fun. I was also lucky to spend an amazing day with Paolo Conti, whose knowledge of and driving ability in Carrara are unrivaled. And without John Logan, William Wallace, and Roy Kligfield sharing their knowledge of Carrara marble and Michelangelo, I would still be trying to understand what I saw there and what Michelangelo did with the stone.

I think I talked to more people about slate than I did about any other stone. Three stand out for their patience, vast depth of knowledge, and enthusiasm: Laurel Grabel, Jack Epstein, and Jeri Jones.

In Italy I was fortunate that Fabrizio Mariotti showed me his family mill and quarry. He also gave me my favorite rock. Eric Doehne and Michael Palladino helped me see the Getty on a human scale. Travertine has had a long history, both geologically and culturally, and what I know of it comes from Robert Folk and Marie Jackson. I was only able to tell the story of NASA and nannobacteria because of Chris Romanek.

Many friends and family put me up and put up with me while I traveled. Thanks to Mike Buckley and Megan Kelso, Nancy and Ira Horowitz, Tim and Amy Johnson-Grass, Niki Lamberg and Adam Shyevitch, Bob and Carol Levine, Janet Protas, and Ruth Schneider.

Several friends read chapters and made helpful suggestions: Megan

Kelso, Jeff Moline, Andy Nettell, Scott Pierce, Jenny Schwarz, Peter Stekel, Scott and Muff Wanek, Irene Wanner, and David Weld. I appreciate their honesty and support.

I interviewed and talked to dozens of people, some of whom appear and many who don't appear in the book. All of those discussions and correspondence helped make up the conglomerate of this book. Thank you to Daniel Abrahamson, Tony Angell, David Barbeau, Kevin Barto, Jim Blachowicz, Amy Brier, Kathleen Burnham, Henry Chafetz, Kent Condie, Dennis Copeland, Steve Cummings, Karen D'Arcy, Kelly Dixon, Mihai Ducea, Peter Dryzewiecki, Bob Emick, Dale Enochs, Lori East, Clayton Fant, Tom Farrell, Tim Fisher, Bruce Fouke, Tom Gaston, Elspeth Gordon, Kirk Johnson, Doug Jones, Brenda Kirkland, Lynne Lancaster, Peter LeTourneau, David Leverington, Greg McHone, Doug McNeill, Virginia Miller, Paula Noble, Irvy Quitmyer, Emma Rainforth, Tony Randozzo, Mark Schmitz, Dale Setterholm, Jim Stagner, Kevin Stewart, Bly Straube, Basil Tikoff, Charles Tingley, Eleanor Toews, Alex Vardamis, and Kate Wellspring.

When I first started to work on this book, I was lucky enough to team up with my agent, Brettne Bloom, whose help in molding my proposal was essential. Jackie Johnson was a patient, thoughtful, and insightful editor. She asked the right questions, found inconsistencies, and tightened my sometimes abstruse science.

I also benefited from the support, enthusiasm, and guidance of Ivan and Carol Doig and David Laskin. I cannot thank you enough for the many years of friendship and inspiration.

And finally, through many walks and talks, some tears, and lots of shots of tequila, I couldn't have written this book without my wife, Marjorie Kittle, and her support, humor, love, and editing.

# Glossary

**architrave** Lowest of three parts of an entablature resting directly on a column.

**balustrade** A row of slender upright posts, or columns, supporting a railing.

**batholith** A large body (at least forty square miles in area) of magma that has cooled underground.

**bedding** Layers or beds of rock; generally applies to sedimentary rock.

**boom** A long beam extending out and usually up from the central pole of a derrick; used for guiding and supporting items.

**brownstone** Sandstone that has a small percentage of iron that has oxidized and colored the rock red to brown. A building, generally a row house, built with brownstone.

**calcite** A mineral made of calcium carbonate, $CaCO_3$.

**Cambrian** A period of geologic time from 542 to 488 million years ago.

**case-hardened** To harden the outside surface. In quarrying this is usually done by letting a stone sit out, or season, for weeks to months.

**chert** A very fine-grained sedimentary rock made of silica.

**conglomerate** A sedimentary rock consisting of rounded sediments of varying sizes.

**console** An ornamented bracket with parallel sides and often topped by a horizontal slab.

**Corinthian** A type of column with an elaborate capital, often depicting acanthus leaves.

**cornice** Projecting ornamental molding that crowns a building or other part of a building, such as an arch or wall.

**Cretaceous** A period of geologic time from 145 to 65 million years ago.

**curtain wall** A nonload-bearing wall built in front of a structure.

**cycad** A group of seed-producing plants with stout trunks and large compound leaves; common in the Jurassic but now much less widespread.

**derrick** A machine for moving heavy objects and often consisting of a mast with a boom attached at or near the base; supporting wires and pulleys allow movement of the boom. The name comes from a seventeenth-century hangman who plied his trade in London.

**diatom** Microscopic, single-celled algae with cell walls made of silica; they live in fresh and saltwater.

**diorite** A plutonic (magma cooled within the earth) rock richer in darker minerals than granite.

**Doric** A column with a plain capital and no base.

**entablature** The part of an order above the column and consisting of the architrave, frieze, and cornice.

**erratic** A large boulder transported and deposited by a glacier.

**extension** In geology this refers to spreading or pulling apart.

**fanlight** A fan-shaped window above a doorway. Can also refer to any shaped window directly atop a door.

**Farallon Plate** A very large oceanic plate that subducted under North America. Its remaining remnants are the Cocos, Rivera, and Juan de Fuca plates.

**feather** A metal shim, often thin and curved, either slightly or at a right angle, at the top.

**floodplain** Area where water spreads when a river floods. Over time the floodplain accumulates sediment.

**foraminifera** Single-celled, generally microscopic protists with shells. They evolved in the Cambrian and are still widespread in marine ecosystems.

**frieze** A decorated or plain band on a wall below the cornice.

**gang saw** A saw made of multiple thin slabs of metal, which cut through stone like a giant bread slicer; named for its gang of blades.

**garret** A room or apartment in the uppermost or attic floor of a house.

**gneiss** High-grade metamorphic rock often consisting of bands of dark and light minerals.

**Gondwana** An ancient supercontinent, roughly 650 to 130 million years old, consisting of Antarctica, Africa, South America, India, Madagascar, and Australia. Also called Gondwanaland. The name refers to a region in India with extensive sedimentary rocks, originally used to characterize other parts of the supercontinent.

**graywacke** A type of sandstone with mud and angular particles of quartz and feldspar.

**hood molding** A projecting molding that blocks rain on a wall or over a window, door, or arch.

**hornblende** A dark mineral common in igneous and metamorphic rocks.

**hot spot** A stationary and localized region of heat in the upper mantle, which generates volcanic activity such as occurs in Hawaii and Yellowstone National Park; also called a mantle plume.

**igneous** Rock that started as a liquid and solidified. Igneous rocks can be intrusive (plutonic) or extrusive (volcanic).

**impost** The part of a pillar or column (usually molding) upon which an arch rests.

**interfinger** Lateral intersecting of layers of different rock types.

**Ionic** A column with a capital consisting of scrolls or volutes on either side of the column.

**keystone** The central stone of an arch or rib.

**lathe** A machine for turning or trimming stone and other substances. The stone is held horizontally and spins against the cutting blade.

**Laurasia** A supercontinent consisting of Siberia, Europe, North America, and parts of China.

**lintel** A horizontal beam, often made of stone, spanning an opening, such as a window or door.

**magma** Underground molten rock.

**mantle** The seventeen-hundred-mile-thick layer between the earth's crust and the core. Plates of the earth's crust move along the mantle's upper layer, called the asthenosphere.

**metamorphic** A rock that has undergone a solid state change due to pressure and temperature.

**mica** Any of a group of transparent, sheetlike minerals including biotite (clear) and muscovite (black).

**moraine** Rocks and sediment transported and deposited by glacial action.

**mudstone** A generic term for a rock consisting of grains generally less than 1/64 of an inch wide.

**nematode** Mostly microscopic multicellular animals also known as roundworms.

**olivine** An olive green mineral that solidifies at high temperatures.

**paleomagnetism** Faint magnetic orientation of iron-rich magnetite crystals preserved within a rock. Used by geologists to ascertain the rock's latitude at time of deposition or crystallization.

**Pangaea** The great supercontinent made of all the continents. It existed between about 300 and 200 million years ago.

**pediment** A low-pitched gable over a portico, window, doors, or façade.

**peridotite** A very dark rock composed almost entirely of pyroxene and olivine. The mantle is primarily made of peridotite.

**planer** A machine that smoothes and/or thins stone and other substances.

**plug** A wedge inserted between two shims, or feathers, and hammered down into a hole to split rock.

**pluton** A body of rock that has solidified within the earth.

**porphyry** A rock with a texture consisting of larger grains set in a fine-grained matrix.

**portico** A roofed space at the front of a building consisting of columns and a pediment.

**protozoa** A single-celled animal, such as an amoeba, a paramecium, and giardia.

**pyroclastic** A general term to describe material ejected violently from a volcano.

**quoins** Dressed stones at the corner of a building.

**repoint** To replace damaged mortar in a wall.

**rift valley** A large-scale valley formed when a tectonic plate or plates begin to split apart.

**Rodinia** A supercontinent made of all the continents, (though they did not look exactly as they do at present), which formed about 1,200 million years ago.

**sedimentary rock** A rock formed by the deposition of sediment by wind, water, or ice; such rocks are usually layered or bedded.

**sleeper** A piece of stone or wood that supports the rails of a railway.

**spall** To break or split off.

**span of horses** A pair of horses.

**stoop** An uncovered platform that rises via stairs to the entrance of a house or other building.

**stucco** A type of plaster made of lime, sand, and fine-grained material mixed with water.

**terrane** A fault-bounded body of rock, with limited extent, characterized by a geologic history different than that of adjacent rocks.

**tonalite** A type of igneous rock also known as quartz-diorite.

**veneer** A thin surface coating on a building. It provides no structural support.

**voussoir** A wedge-shaped stone forming one part of an arch.

**yoke of oxen** A pair of oxen.

# Notes

## 1: "THE MOST HIDEOUS STONE EVER QUARRIED"

1. Junius Henri Browne, *Great Metropolis; A Mirror of New York* (Hartford: American Publishing Co., 1869), 222.
2. James Richardson, "The New Homes of New York: A Study of Flats," *Scribner's Monthly* 8, no. 1 (1874): 67.
3. Marianna G. van Rensselaer, "Recent Architecture in America V: City Dwellings," *Century Illustrated Magazine* 31, no. 4 (1886): 550.
4. Details on Vanderbilt, his money, and his building come from Robert A. M. Stern, Thomas Mellins, and David Fishman, *New York 1880: Architecture and Urbanism in the Gilded Age* (New York: Monacelli Press, 1999), 568–97.
5. Edith Wharton, *A Backward Glance* (New York: D. Appleton-Century Company, 1934), 55.
6. Iron makes up about 3 percent of the rock, so it would not work to mine it.
7. Browne, *Great Metropolis*, 222.
8. Despite, or in spite of, its bad reputation, perhaps the most famous brownstone denizens of modern times moved into their basement apartment on November 10, 1969. As such, *Sesame Street*'s Bert and Ernie were catalysts for reintroducing brownstones to the greater world. And who could be a better spokescurmudgeon for the stoop than Oscar the Grouch, who lived in his garbage can next to Bert and Ernie's entryway.
9. Alex Barrett, interview with author, New York, NY, October 23, 2006.
10. The bones of *Anchisaurus* were the first dinosaur parts found in North America. Solomon Edwards discovered the bones in 1818 when digging a well in East Windsor, Connecticut. Dr. Nathan Smith, author of the first scientific description of the fossils, wrote, "Whether they are human or brute animal bones, it is an important fact as it relates to Geology." "Fossil Bones found in red sand stone," *American Journal of Science* 2, no. 1 (1820): 146–47. Nicholas G.

McDonald designated these fossils as the first found in "Connecticut in the Age of Dinosaurs," *Rocks & Minerals* 70 (1995): 412–18.

11. There are numerous accounts of the Noah's Raven discovery and its eventual acquisition by Hitchcock. The most thorough comes from Nancy Pick, *Curious Footprints: Professor Hitchcock's Dinosaur Tracks & Other Natural History Treasures at Amherst College* (Amherst, MA: Amherst College Press, 2006). Pick's book also contains good information on Hitchcock and his life.
12. Information on locations and habits of *Anomoepus* comes from Paul E. Olsen and Emma C. Rainforth, "The Early Jurassic Ornithischian Dinosaurian Ichnogenus Anomoepus" in *The Great Rift Valleys of Pangea in Eastern North America: Sedimentology, Stratigraphy, and Paleontology* vol. 2, edited by Peter M. Letourneau and Paul E. Olsen (New York: Columbia University Press, 2003).
13. Steve Sauter, interview with author, Amherst, Massachusetts, October 26, 2006. I also based much of my description of Hitchcock on my conversation with Sauter.
14. Pick, *Curious Footprints,* 6.
15. Edward Hitchcock, "Ornithichnology. Description of the Foot Marks of Birds, (Ornithichnites) on new Red Sandstone in Massachusetts," *American Journal of Science and Arts* 29, no. 2 (1836): 307–41.
16. Edward Hitchcock, *Reminiscences of Amherst College, Historical, Scientific, Biographical and Autobiographical: Also, of Other and Wider Life Experiences*, (Northampton, MA: Bridgman & Childs, 1863), 84.
17. Ibid., 85.
18. Hitchcock used the phrase "gem of the Cabinet" for a species he called *Brontozoum Sillimanium* in *Ichnology of New England* (Boston: William White, 1858), 68.
19. Alison Guinness, interview with author, Portland, Connecticut, October 24, 2006.
20. Many of the statistics and quotes on the history of brownstone come from two papers by Alison Guinness. Alison Guinness, "Heart of Stone: The Brownstone Industry of Portland, Connecticut," in *The Great Rift Valleys of Pangea,* 224–47. Alison Guinness, "The Portland Brownstone Quarries," *The Chronicle of the Early American Industries Association, Inc.* 55, no. 3 (2002): 95–112.
21. J. S. Bayne, *Town of Portland: History of Middlesex County, Connecticut* (New York: J. B. Beers & Co., 1884), 516.
22. Such rock is known as freestone; the term applies to any sandstone or limestone that cuts easily. Freestone predates both of the more specific geologic terms by several hundred years.
23. Edmund M. Blunt, *Blunt's Stranger's Guide to the City of New-York* (New York: Edmund M. Blunt, 1817), 45.
24. James Fenimore Cooper didn't agree with the builders and thought the deep red brownstone was in "far better taste" than the marble front. "The moment the rear of the City Hall is seen, I was struck with an impression of the mag-

nificent effect which might be produced by the use of its material in Gothic architecture," he wrote in letter VIII to Baron Von Kemperfelt in *Notions of the Americans: Picked up a Traveling Bachelor* (Philadelphia: Carey, Lea & Care, 1835).

25. Alain de Botton, *The Architecture of Happiness* (New York: Pantheon Books, 2006), 28.
26. Charles Lockwood, *Bricks and Brownstone* (New York: McGraw-Hill, 1972), 104.
27. A. J. Downing, *The Architecture of Country Houses* (New York: D. Appleton & Co., 1850), 198, 200.
28. Bayne, *Town of Portland*, 520.
29. All statistics from articles by Alison Guinness. The building now houses the exclusive Pacific Union Club, which acquired the structure in 1907 from Flood's daughter. After the earthquake, the club gutted and rebuilt the house with stone from the Portland quarry.
30. Lewis Mumford, *The Brown Decades: A Study of the Arts in America, 1865–1895* (New York: Harcourt, Brace and Company, 1932), 2–3.
31. Mike Meehan, interview with author, Portland, Connecticut, October 25, 2006.

## 2: THE GRANITE CITY

1. William S. Pattee, *A History of Old Braintree and Quincy* (Quincy: Green & Prescott, 1878), 498.
2. Shaw gave this speech on December 13, 1859, at the 473rd meeting of the American Academy of Arts and Sciences in Cambridge, Massachusetts. The speech was reprinted in volume 4 of the *Proceedings of the American Academy of Arts and Sciences*, 353–59.
3. Shaw's speech is the only known reference to Mr. Tarbox. He doesn't appear in newspapers, death records, or any other official document from the period, at least none James and Mary Gage could find. They are coauthors of *The Art of Splitting Stone: Early Rock Quarrying Methods in Pre-Industrial New England* (Amesbury, MA: Powwow River Books, 2002). No one knows exactly what Tarbox said. I based my description of the Tarbox method on the Gages' book, Shaw's speech, and on an interview in Boston, Massachusetts, with James Gage in December 2003.
4. Pattee, *History of Old Braintree*, 515.
5. Bartram's letter is quoted in the *Art of Splitting Stone*, p. 26.
6. Information on the Finnish and Egyptian methods comes from George P. Merrill, *Stones for Building and Decoration*, 3rd ed. (New York: John Wiley & Sons, 1903), 393. Merrill's is one of the best books detailing the stone industry.
7. Smoky quartz can be turned clear again by heat. I have fond memories of fun times in college when we took a crystal of smoky quartz, placed it on a stove, cranked the burner up, and drove out the color. Those were the days!

8. Arthur W. Brayley, *History of the Granite Industry of New England* (Boston: National Association of Granite Industries of the United States, 1913), 22–23.
9. Willard left no record of his trek; the story comes from a single sentence written on a blank page in the notes of the Bunker Hill Building Committee in August 1849. The author was Amos Lawrence, a central player in the construction of the monument since the beginning. Willard's journey is mentioned in William Wheildon's *Memoir of Solomon Willard* (Boston: Bunker Hill Monument Association, 1865), 108.
10. George Washington Warren, *The History of Bunker Hill Monument Association During the First Century of the United States of America* (Boston: James R. Osgood and Company, 1877), 202.
11. Ibid., 47.
12. Ibid., 157.
13. Quoted in Sarah J. Purcell, "Commemoration, Public Art, and the Changing Meaning of the Bunker Hill Monument," *Public Historian* 25, no. 2 (2003): 55–71.
14. Wheildon, *Memoir*, 109.
15. Warren, *History*, 140.
16. Information and quotes from Bryant come from Charles B. Stuart, *Lives and Works of Civil and Military Engineers of America* (New York: D. Von Nostrand, 1871), 119–31.
17. Information on Perkins's contributions to the project comes from Carl Seaburg and Stanley Paterson, *Merchant Prince of Boston: Colonel T. H. Perkins, 1764–1854* (Cambridge, MA: Harvard University Press, 1971).
18. Ibid., 331, 333–34.
19. Vic Campbell, interview with author, Quincy, Massachusetts, January 20, 2006.
20. Fred Gamst, "The Context and Significance of America's First Railroad, on Boston's Beacon Hill," *Technology and Culture* 33, no. 1 (1992): 66–100.
21. Seaburg and Paterson, *Merchant Prince*, 337–38.
22. Richard Bailey, interview with author, Boston, Massachusetts, January 19, 2006.
23. Percy E. Raymond, "Notes on the Ontogeny of Paradoxides, with a description of a new species from Braintree, Mass," *Bulletin of the Museum of Comparative Zoology* 58, no. 4 (1914): 225–47.
24. W. O. Crosby and G. F. Loughlin, "A Descriptive Catalogue of the Building Stones of Boston and Vicinity," *Technology Quarterly* 17 (1904): 165–85.
25. *The Poetical Works of Oliver Wendell Holmes* (Boston: Houghton, Mifflin and Company, 1892), 19.
26. Wheildon, *Memoir*, 142.
27. Warren, *History*, 215.
28. Ibid., 216.
29. Ibid., 280–83.
30. Ibid., 284, 298.

31. S. Willard, *Plans and Sections of the Obelisk on Bunker's Hill* (Boston: Chas. Cook's Lith., 1843), 9.
32. Jane Holtz Kay, *Lost Boston* (Boston: Houghton Mifflin, 1980), 130.

## 3: POETRY IN STONE

1. "Winged Rock," *The Selected Poetry of Robinson Jeffers* (New York: Random House, 1937), 361.
2. *The Selected Letters of Robinson Jeffers, 1897–1962*, ed. Ann N. Ridgeway (Baltimore: Johns Hopkins University Press, 1968), 213.
3. Stewart Brand, *How Buildings Learn: What Happens After They're Built* (New York: Viking, 1994), 49.
4. "Tor House," *Selected Poetry*, 197.
5. Aaron Yoshinobu, interview with author, Carmel, California, April 28, 2006.
6. "To the House," *Selected Poetry*, 82.
7. "Tamar," Ibid., 49.
8. "The Old Stonemason," *Robinson Jeffers Selected Poems*, the centenary edition, ed. Colin Falck (Manchester, England: Carcanet, 1987), 81.
9. Ibid.
10. "Gray Weather," *Selected Poetry*, 572.
11. "The Inhumanist," Robinson Jeffers, *The Double Axe and Other Poems* (New York: Random House, 1948), 54.
12. "Woman at Point Sur," *The Collected Poetry of Robinson Jeffers*, vol. 1, ed. Tim Hunt (Palo Alto, CA: Stanford University Press, 1988), 309.
13. "The Inhumanist," *Double Axe*, 57.
14. Melba Berry Bennett, *The Stone Mason of Tor House* (Los Angeles: Ritchie, 1966), 4.
15. *Selected Letters*, 353.
16. Bennett, *Stone Mason*, 24.
17. Sidney S. Albert, *A Bibliography of the Works of Robinson Jeffers* (New York: Bert Franklin, 1968), XVI.
18. Bennett, *Stone Mason*, 31.
19. Ibid., 47.
20. *Selected Letters*, 353.
21. *Los Angeles Times* review, December 8, 1912, written under the name Willard Huntington Wright but generally attributed to Jeffers according to Alex Vardamis, *The Critical Reputation of Robinson Jeffers; A Biographical Study* (Hamden, CT: Archon Books, 1972), 35.
22. Lawrence Clark Powell, *An Introduction to Robinson Jeffers* (Ph.D. diss., University of Dijon, 1932), 9.
23. Bennett, *Stone Mason*, 71.
24. Ibid., 70.
25. Robert Brophy, "M. J. Murphy Masterbuilder and Tor House," *Robinson Jeffers Newsletter* 78 (1990): 24–27.

26. "The Bed by the Window," *Selected Poetry*, 362.
27. "The Last Conservative," *The Collected Poetry of Robinson Jeffers*, vol. 3, ed. Tim Hunt (Palo Alto, CA: Stanford University Press, 1991), 18.
28. Dennis Copeland, archivist with the Monterey Public Library, wrote in an e-mail in April 2006 that the salvaged porthole most likely came from the brig *Natalia*, which had sunk near Monterey on December 21, 1834, and not from Napoleon's ship the *Inconstant*, as reported by Donnan Jeffers in *The Stone of Tor House*.
29. Dave Barbeau, phone interview with author, April 2006.
30. About 27 million years ago the subducting Farallon Plate disappeared completely under North America. The Pacific Plate, which was moving northwest, became dominant and began to carry material up the California coast, and the San Andreas Fault was born.
31. James M. Mattinson and Eric W. James, "Salinian Block U/Pb Age and Isotopic Variations: Implications for Origin and Emplacement of the Salinian Terrane," *Tectonostratigraphic Terranes of the Circum-Pacific Region: Circum-Pacific Council Energy Mineral Resources*, ed. D. G. Howell, Earth Science, series 1 (1985): 215–26.
32. Dana Gioia, review of *Rock and Hawk: A Selection of Shorter Poems by Robinson Jeffers*, ed. Robert Hass, *Nation* 246, no. 2 (January 16, 1988): 56–64.
33. "Old age hath clawed me," *The Collected Poetry of Robinson Jeffers*, vol. 4, ed. Tim Hunt (Palo Alto, CA: Stanford University Press, 2000), 484.
34. "Flight of Swans," *Selected Poetry*, 577.
35. "Pelicans," *Collected Poetry*, v. 1, 207.
36. Ibid., 207.
37. "Birds and Fishes," *Collected Poetry*, v. 3, 426.
38. "The Loving Shepherdess," *Selected Poetry*, 357.
39. "The Women at Point Sur," *Selected Poetry*, 154.
40. "All night long," *Collected Poetry*, v. 3, 481.
41. "Pelicans," *Collected Poetry*, v. 1, 207.
42. "Night Without Sleep," *Selected Poetry*, 609.
43. "Rock and Hawk," *Selected Poetry*, 563.
44. "The Last Conservative," *Collected Poetry*, v. 3, 418.
45. "Tor House," *Selected Poetry*, 197.

## 4: DEEP TIME IN MINNESOTA

1. Siccar appears to derive from scaur, a rock or precipice.
2. Basil Tikoff, phone interview with author, November 2005. Tikoff is also the "one geologist" quoted at the end of the chapter.
3. J. C. Beltrami, *Pilgrimage in Europe and America, leading to the discovery of the sources of the Mississippi and Bloody river; with a description of the whole course of the former and of the Ohio* (London: Hunt and Clarke, 1828), 318.
4. Ibid., 319.

5. George Thiel and Carl Dutton, "Architectural, Structural, and Monumental Stones of Minnesota," *Minnesota Geological Survey Bulletin* 25 (1935): 92.
6. William Keating, *Narrative of an Expedition to the Source of St. Peter's River, Lake Winnepeek, Lake of the Woods, &c. performed in the Year 1823* (Philadelphia: George B. Whittaker, 1824), 350.
7. Ussher (1581–1656) was not alone in generating dates. In 1809 Irish professor William Hales listed 156 proposed Creation dates ranging from sixty-five hundred to thirty-six hundred years before Christ's birth.
8. John Playfair, "Biographical Account of the Late Dr. James Hutton," *Transactions of the Royal Society of Edinburgh* 5 (1805).
9. Most information on age of Earth dates comes from *The Age of the Earth: from 4004 BC to AD 2002*, ed. C. L. E. Lewis and S. J. Knell, Geological Society Special Publication 190 (2001).
10. S. S. Goldich, A. O. C. Nier, H. Baadsgaard, and J. H. Hoffman, "$K^{40}/A^{40}$ dating of Precambrian rocks of Minnesota (abs)," *Geological Society of America Bulletin* 67 (1956): 1698–99.
11. Pb is the chemical symbol for lead because its Latin name was *plumbum nigrum*, black lead, which also gave us the word plumbing. Uranium received its name from its discoverer, Martin Klaproth, who named the new element in 1789 for the then most recently discovered planet, Uranus, which referred to Urania, the muse of astronomy and geometry. Coincidentally, Klaproth also discovered zirconium.
12. Zircon derives from the Persian words *zar*, or gold, and *gun*, or color, in reference to the mineral's color.
13. Typical magmas are rich in a stew of various elements, most of which occur in miniscule amounts. When minerals crystallize they can incorporate these trace elements, although the elements don't show up in the mineral's chemical formula. For example, zircon's formula is $ZrSiO_4$ but it can also include promethium, hafnium, yttrium, and samarium, as well as uranium.
14. Pat Bickford, phone interviews with author, June 2007.
15. E. J. Catanzaro, "Zircon ages in southwestern Minnesota," *Journal of Geophysical Research* 68 (1963): 2045–48.
16. S. S. Goldich and C. E. Hedge, "3,800-Myr granitic gneisses in south-western Minnesota," *Nature* 252 (December 6, 1974): 467–68.
17. M. E. Bickford, J. L. Wooden, and R. L. Bauer, "SHRIMP study of zircons from Early Archean Rocks in the Minnesota River Valley: Implications for the tectonic history of the Superior Province," *Geological Society of America Bulletin* 118 (2006), 94–108.
18. Robert Stern, phone interview with author, May 2007.
19. Kent Condie, phone interview with author, June 2007.
20. Mark Gross, interview with author, Morton, Minnesota, May 21, 2007.
21. D. L. Southwick and V. W. Chandler, "Block and shear-zone architecture of the Minnesota River Valley subprovince: implications for late Archean accretionary tectonics," *Canadian Journal of Earth Science* 33 (1996): 831–47.

22. You also may see the term migmatite, a word used by geologists to describe well-mixed rocks such as the Morton. Not all gneisses are migmatic but all migmatites are gneiss.
23. Dan Rea, interview with author, Cold Spring, Minnesota, May 21, 2007.
24. G. W. Featherstonhaugh, *A Canoe Voyage up the Minnay Sotor* (London: Richard Bentley, 1847), 326.
25. Named for Civil War veteran General Gouverneur Kemble Warren, who hypothesized such a river during a survey in 1868.
26. This is the total area ever covered by the lake. Its maximum size at one time was 324,000 square miles, about the combined size of Texas and Oklahoma. Data comes from two papers. David Leverington, Jason Mann, and James Teller, "Changes in the Bathymetry and Volume of Glacial Lake Agassiz between 9200 and 7700 $^{14}C$ yr B.P.," *Quaternary Research* 57 (2002): 244–52, and James Teller and David Leverington, "Glacial Lake Agassiz: A 5000 yr history of change and its relationship to the $^{18}O$ record of Greenland," *GSA Bulletin* 116 (2004): 729–42.

## 5: THE CLAM THAT CHANGED THE WORLD

1. Robert M. Weir, "Charles Town Circa 1702: On the Cusp," *El Escribano* 39 (2002): 64–79.
2. Verner Winslow Crane, *The Southern Frontier* (Ann Arbor: University of Michigan Press, 1956), 75.
3. Details from the attack come from Charles W. Arnande, "The Siege of St. Augustine in 1702," *University of Florida Monographs*, Social Sciences, no. 3, Summer 1959.
4. Joe Brehm, interview with author, St. Augustine, Florida, January 5, 2007.
5. The peaceful transfer was part of the Adams-Onis Treaty, which settled territorial boundaries between Spain and the United States in Florida, Texas, and the Rocky Mountains.
6. Joe Brehm told me that in 1935, after Errol Flynn and Olivia de Havilland starred in the movie *Captain Blood*, which featured a fortress with a moat, the National Park Service received hundreds of pounds of mail requesting that the fort fill its moat.
7. Gastroliths are rocks swallowed by birds that aid in digestion. Dinosaurs also swallowed gastroliths.
8. Wendy B. Zomlefer and David E. Giannasi, "Floristic Survey of Castillo de San Marcos National Monument, St. Augustine, Florida," *Castanea* 70 (2005): 222–36.
9. Michael Gannon, *The New History of Florida* (Gainesville: University Press of Florida, 1996), 30.
10. The best information on forts and life in St. Augustine comes from Verne E. Chatelain, "The Defenses of Spanish Florida 1565–1763," *Carnegie Institution of Washington Publication* 511 (1941). Other good information comes from Jeannette Connor, "The Nine Old Wooden Forts of St. Augustine," parts 1–2, *Florida Historical Society Quarterly* IV (1967): 103–11, 171–80; Eugene Lyon, "The First

Three Wooden Forts of Spanish St. Augustine," *El Escribano* 34 (1997): 140–57; Albert Manucy, "Building Materials in 16th-Century St. Augustine," *El Escribano* 20 (1984): 51–71.

11. Connor, "Nine Old Wooden Forts," 110.
12. Michael Gannon first coined this concept in his *Florida, A Short History* (Gainesville: University Press of Florida, 2003), when he wrote that "by the time the Pilgrims came ashore at Plymouth, St. Augustine was up for urban renewal." Elsbeth Gordon added to Gannon's phrase in her fascinating *Florida's Colonial Architectural Heritage* (Gainesville: University Press of Florida, 2002). Floridians take deep umbrage at the prevailing idea that America's colonial story begins in Virginia and Massachusetts and not in Florida.
13. Connor, "Nine Old Wooden Forts," 172.
14. Ibid., 172.
15. The best description of Searles's attack occurs in Luis Rafael Arana, "The Basis of a Permanent Fortification," in *El Escribano* 36 (1999), 3–11.
16. Arana describes this maneuvering in detail, from *El Escribano* 36.
17. *The History of Castillo de San Marcos* (St. Augustine: Historic Print & Map Co., 2005).
18. An article in the 1950s described thousands of goldfish swimming in abandoned coquina quarries. People would travel to the island to "pan gold."
19. Olaf Ellers, phone interview with author, December 2006.
20. Marjorie Kinnan Rawlings, *Cross Creek Cookery* (New York: Fireside, 1996), 16–17.
21. Donald F. McNeill, "Petrologic Characteristics of the Pleistocene Anastasia Formation, Florida East Coast" (master's thesis, University of Florida, 1983).
22. Description of calcite, aragonite, formation, and seasoning based on author's phone conversation with Donald McNeill, January 2007.
23. Based on author's conversation with Roger Portell, Division of Paleontology, Florida Museum of Natural History, who kindly showed me the ghost crabs and talked in depth about the deposition of the Anastasia Formation, Gainesville, Florida, January 2007.
24. Description of life of *Donax* based on phone conversation with Olaf Ellers. For more information consult his papers: "Biological Control of Swash-Riding, in the Clam *Donax variabilis*," *Biological Bulletin* 189 (1995): 120–27, and "Discrimination Among Wave-Generated Sounds by a Swash-riding Clam," *Biological Bulletin* 189 (1995): 128–37.
25. Albert Manucy, *The Houses of St. Augustine* (St. Augustine: St. Augustine Historical Society, 1962).
26. The following section on the construction of the castillo was pieced together from *The History of Castillo de San Marcos* and "The Defenses of Spanish Florida 1565–1763."
27. *History of Castillo*, 28.
28. Tabby, also called *piedra de ostion* (oyster stone), was made by mixing water and sand with lime and oyster shells. The shells came from ancient Native American

middens. Tabby was used primarily in the 1700s in buildings as far north as Charleston, South Carolina.

29. George Fairbanks, *The Spaniards in Florida: Comprising the Notable Settlement of the Huguenots in 1564 and the History and Antiquities of St. Augustine* (Jacksonville, FL: Columbus Drew, 1868), 103.
30. Kathyrn Hall Proby, *Audubon in Florida, with Selections from the Writings of John James Audubon* (Coral Gables, FL: University of Miami Press, 1974), 15.
31. *History of Castillo*, 43.
32. British rule provides the framework for one of the few novels set in St. Augustine. Eugenia Price's *Maria* is a fictionalized account of Mary Evans, who arrived in 1763 and became quite wealthy before losing her money in a May-December marriage. She died in 1792. Neither particularly good nor bad, it does not, however, mention coquina. St. Augustine was also the setting for one movie, an absolutely horrible attempt at comedy called *Illegally Yours*. It does show the castillo, which has become the mansion of a rich oddball, who has decorated it with Greek statues and a Chinese pagoda. Gary Cooper also starred as a rebel soldier in a movie that shows the castillo, this time named Fort Infanta. *Distant Drums* focuses on the Seminole wars and features Cooper and a handful of men taking the fort. Again no one notes the coquina but one of Cooper's men does note that the fort was designed by Enrico Garcia, one of Spain's greatest military architects, and that it is impossible to take with less than a brigade.
33. Kathleen Deagan, "A New Florida & A New Century: The Impact of the English Invasion on Daily Life in St. Augustine," *El Esribano* 39 (2002): 102–12.
34. Alfred J. Morrison, ed., *Travels in the Confederation, 1783–1784. From the German of Johann David Schoepf* (New York: Bergman, 1968), 250.
35. Proby, *Audubon in Florida*, 17.
36. Oddly, Audubon didn't paint the fort. George Lehman, a Swiss artist, who accompanied Audubon on his 1831–1832 trip down the east coast of Florida, painted the castillo while they stayed in St. Augustine (January 1831). This was not unusual. Audubon often hired another artist to accompany him on his journeys, to paint plants and landscapes for use as backgrounds in his prints, which allowed Audubon to concentrate on his birds.
37. Harriet Beecher Stowe, *Palmetto-Leaves* (Boston: James Osgood and Co., 1873), 206.
38. Gary Wilson, Lakeview Dirt Co., Inc., phone interview with author, February 2007.

## 6: AMERICA'S BUILDING STONE

1. Keith said that his analogy was to the holy trinity of Cajun cooking—onions, green pepper, and celery—and not to the better-known holy trinity of Father, Son, and Holy Spirit. Interview with author, Bloomington, Indiana, September 10, 2007.

2. If you look at the back of a one-dollar bill, you will also see limestone. The pyramids of Egypt are made of a fossil-rich limestone.
3. Amy Brier, interview with author, Bloomington, Indiana, September 12, 2007.
4. Specifically the species is *Globoendothyra baileyi.*
5. Todd Thompson, interview with author, Bloomington, Indiana, September 10, 2007.
6. Until the late 1980s, scientists thought that stalked crinoids could not move, and until 2005 they had never photographed one moving. That year researchers in the Bahamas videotaped one crawling across the sandy bottom. The animal had broken off its holdfast and pulled itself along by its flexible arms. It moved one to two inches per second and was last seen speeding away from a determined and hungry sea urchin.
7. Merrill, *Stones for Building*, 405.
8. Ibid., 405.
9. Joseph Batchelor, "An Economic History of the Indiana Oolitic Limestone Industry," *Indiana Business Studies Study* 27 (1944): 40.
10. Ibid., 40.
11. George Jones, interview with author, Bedford, Indiana, September 13, 2007. I could not have seen all of the quarries and mills and met their owners without the generous help of Jim Owens, executive director of the Indiana Limestone Institute, the industry's trade association.
12. In past times the colors used to be buff and blue. Originally buyers preferred blue over buff. Blue or gray is the natural color, with water oxidizing the stone to buff.
13. Only one Indiana mill uses the giant bread-slicer-like gang saws described in chapter 4. Bybee Stone has stuck with a few gang saws because they give the stone a rougher, more rustic finish. Their decision paid off in 2001, when they won the contract to repair the Salem Limestone walls damaged by the September 11 bombing of the Pentagon. By June 2002, Bybee had sent 2.1 million pounds of cut limestone to Washington, D.C.
14. Bob Thrasher, interview with author, Bloomington, Indiana, September 11, 2007.
15. Will Bybee, interview with author, Ellettsville, Indiana, September 10, 2007.
16. "The Great Rebuilding," *Chicago Tribune*, October 9, 1872: 8.
17. Ibid.
18. The *Chicago Tribune* (August 20, 1876) called the bidding process "utter absurdity." The board's initial choice of contractor put in a bid of $895,000. When that bid failed, the board, or "the Ring," hoped to make their money by finding an architect who would help plunder the system.
19. Both quotes come from: *Twenty First Annual Report*, Indiana Department of Natural Resources, Indianapolis (1896): 323.
20. Jim O'Connor, "Building Stones of Our Nation's Capital," unpublished manuscript, unknown date, 46.
21. If you want to get a feel for the Indiana stone industry, I recommend the movie *Breaking Away*. It takes place in Bloomington in the late 1970s and centers on

four recent high school graduates, Dave, Moocher, Cyril, and Mike. Ostensibly about the relationship between mill workers, or Cutters, and college kids, *Breaking Away* is filled with the angst and self-doubt of young men who cannot follow their father's footsteps. "They're gonna keep calling us 'Cutters.' To them it's just a dirty word. To me it's just something else I never got a chance to be," says Mike. *Breaking Away* has some fine scenes of mills, abandoned quarries, and biking.

22. Dale Enochs, interview with author, Bloomington, Indiana, September 11, 2007.

## 7: POP ROCKS, PILFERED FOSSILS, AND PHILLIPS PETROLEUM

1. Dorothy Smith, interview with author, Lamar, Colorado, July 10, 2007.
2. Kirk Johnson, phone interview with author, July 2007.
3. In one case, a well-known cartoonist was on vacation near what would become Florissant Fossil Beds National Monument, in Colorado, and saw a piece of petrified wood that he wanted for his home garden. Since the wood was on private land, he was able to buy it. His wife, however, didn't want the fossil at home so it ended up in the cartoonist's amusement park, Disneyland.
4. Lester Ward, "Sketch of Paleobotany," *Fifth Annual Report United States Geological Survey* (1885): 385.
5. From a letter to John Ray, quoted in *Three Physico-Theological Discourses* (New York: Arno Press, 1978), 190.
6. J. B. Delair and W. A. S. Sarjeant, "The Earliest Discoveries of Dinosaurs: The Records Reexamined," *Proceedings of the Geologists' Association* 113 (2002): 185–97.
7. Commentary on Chapter 2, Verse 12 of Genesis, from *Luther's Works, Volume 1 (Lectures on Genesis Chapters 1–5)* (St. Louis: Concordia Publishing House, 1955), 98.
8. Ward, "Sketch," 390–91.
9. When I lived in Moab, Utah, I heard of one of the rarer petrifying agents, uranium. A friend told me of fossilized wood where uranium ore had replaced entire trees, which had been mined limb by limb.
10. Description of formation of petrified wood based on phone interview with George Mustoe, geologist at Western Washington University, August 8, 2007.
11. Granger was not the first to find this cabin. William Reed and Frank Williston had also seen it in the 1870s but they concluded that the bones, which a fellow paleontologist referred to as "head cheese," were too deteriorated for further exploration. The best information on these discoveries comes from two papers: John McIntosh, "The Second Jurassic Dinosaur Rush," *Earth Sciences Monthly* 9, no. 1 (1990): 22–27, and Vincent Morgan and Spencer G. Lucas, "Walter Granger, 1872–1941, Paleontologist," *New Mexico Museum of Natural History and Sciences* 19, (2002).
12. The Texas Tourist Camp in Decatur, Texas, and the Petrified Wood Park in Lemmon, South Dakota, also feature petrified wood structures. Both were built in the 1930s and both are on the National Register of Historic Places.

13. James Agee, "The Great American Roadside," *Fortune* (September 1934): 53–63, 172, 176–77.
14. William Kaszynski, *The American Highway: The History and Culture of Roads in the United States* (Jefferson, NC : McFarland & Company, Inc., 2000), 40–42.
15. John Jakle and Keith A. Sculle, *The Gas Station in America* (Baltimore: Johns Hopkins University Press, 1994), 51.
16. Ibid., 52.
17. Warren James Belasco, *Americans on the Road: From Autocamp to Motel, 1910–1945* (Cambridge: MIT Press, 1979), 35.
18. Michael Karl Witzel, *The American Gas Station* (Osceola, WI: Motorbooks International, 1992), 34.
19. Jakle and Sculle, *Gas Station*, 58.
20. Witzel, *American Gas,* 60.
21. Carolyn Peyton, interview with author, Lamar, Colorado, July 10, 2007.
22. Greg Emick, interview with author, Lamar, Colorado, July 10, 2007.
23. Quoted in *Lamar Daily News*, November 17, 1978.

## 8: THE TROUBLE WITH MICHELANGELO'S FAVORITE STONE

1. All quoted phrases come from Edward Durell Stone, *Evolution of an Architect* (New York: Horizon Press, 1962). They are found on pages 149, 143, 143, and 151, respectively.
2. Ms. Swearingen's quote from 1970 is referred to in an article in the *Chicago Sun-Times*, March 7, 1989.
3. All information on building, examination, and cladding of the Amoco Building is from Ian R. Chin, *Proceedings of the Seminar on the Recladding of the Amoco Building in Chicago, IL Held on November 11, 1993* (Chicago: The Chicago Committee on High Rise Buildings, 1994).
4. John Logan, phone interview with author, November 2007.
5. Chin, *Proceedings*, 4–1.
6. Information based on phone interview with Jon Mendelson and Karen D'Arcy, Division of Science, Governors State University, April 2007.
7. Giorgio Vasari, *Lives of the Artists*, trans. Julia Conaway Bondanella and Peter Bondella (Oxford: Oxford University Press, 1991), 415.
8. Most of the details in the following section comes from work originally collected by William Wallace in his fascinating and thorough *Michelangelo at San Lorenzo: The Genius as Entrepreneur* (Cambridge: Cambridge University Press, 1994). I could not have put together this story of Michelangelo without Dr. Wallace's book and his generosity in answering my numerous questions.
9. Michelangelo to Domenico Buoninsegni, May 2, 1517, *Letters Translated from the Original Tuscan*, ed. E. H. Ramsden (Palo Alto: Stanford University Press, 1963), 105–7.
10. Michelangelo to Domenico Buoninsegni, January 1519, *Letters*, 123.

11. Using the trench and wedge technique, the Egyptians quarried a block 137 feet long weighing 1,168 tons. A flaw in the stone prevented it from being used and it still rests in the quarry at Aswan. By the fifth century BCE, according to Xenophon, the Greeks had developed quarries in Piraeus large enough to hold hundreds of prisoners of war from Syracuse.
12. Michelangelo to Buonarroto di Lodovici Simoni, July 28, 1515, *Letters*, 93.
13. Michelangelo to Derto da Filicaia, August 1518, *Letters*, 117–18.
14. Michelangelo to Pietro Urbano, April 20, 1519, *Letters*, 124–25.
15. Charles Dickens, *Pictures from Italy* (London: William H. Colyer, 1846), 35–36.
16. Michelangelo to Lionardo, December 21, 1518, *Letters*, 121.
17. In February 2007 Gabriele Morolli, an architectural historian in Florence, reported that he had found three of Michelangelo's columns in Pisa. He transported one to Florence and planned to dig up the others. Many art historians remain skeptical of Morolli's find, and no information has been forthcoming since the discovery.
18. Michelangelo to unknown, possibly Ser Bonaventura di Leonardo, March 1520, *Letters*, 128, 130–31.
19. The great classics archaeologist John Ward-Perkins used this phrase.
20. Rodolfo Lanciani, *Ancient Rome in the Light of Recent Discoveries* (Boston: Houghton Mifflin, 1891), 243.
21. J. Clayton Fant, "The Roman Emperors in the Marble Business: Capitalists, Middlemen or Philanthropists," in *Classical Marble: Geochemistry, Technology, Trade*, ed. Norman Herz and Marc Waelkens (Dordrect, Netherlands: Kluwer Academic Publishers, 1988), 147–59.
22. "This marble is known to Roman stone-cutters as Marmo Greco Fetido (fetid Greek marble) and Marmo cipolla (onion marble), because when sawn it emits a fetid odour": Mary Winearls Porter, *What Rome Was Built With: A Description of the Stones Employed in Ancient Times for its Building and Decoration* (London: Henry Frowde, 1907), 77. This is not an unusual phenomenon; organic remains in the rock can disintegrate and form a sulfurous gas, which gets trapped in the crystal lattice. Breaking the stone releases the gas.
23. Giuseppe Bruschi, Antonio Criscuolo, Emanuela Paribeni, and Giovanni Zanchetta, "$^{14}$C-dating from an old quarry waste dump of Carrara marble (Italy): evidence of pre-Roman exploitation," *Journal of Cultural Heritage* 5 (2004): 3–6. This early date has not been widely accepted.
24. During the first excavation of Marmorata, in 1868, workers found over twelve hundred large blocks of marble and several thousand cut slabs. Pope Pius XII controlled the dig, which has been called an appalling operation because of the lack of detailed records. The pope used the ancient stones to rebuild churches throughout Italy and to reset the marble paving around the Pantheon. He also sent material across Europe and as far away as Argentina. Clayton Fant argues convincingly that much of the Marmorata material was substandard or defective and therefore rejected, as opposed to other archaeologists who argue that overproduction and oversupply led to abandonment of the stone at Marmorata.

25. Jerry Brotton, *The Renaissance Bazaar: From Silk Road to Michelangelo* (Oxford: Oxford University Press, 2003), 106.
26. James Ackerman, *The Architecture of Michelangelo* (London: A. Zwemmer, 1961), 139.
27. *Reactions to the Master: Michelangelo's Effect on Art and Artists in the Sixteenth Century*, ed. Francis Ames-Lewis and Paul Joannides (Burlington, VT: Ashgate Publishing, 2003), 143.
28. Ackerman, *Architecture*, 6.
29. Gloria Ciarapica and Leonsevero Passeri, "Late Triassic and Early Jurassic Sedimentary Evolution of the Northern Apennines: An Overview," *Bollettino della Societa Geologica Italiana* 124 (2005): 189–201
30. Leonsevero Passeri and Federico Venturi, "Timing and Causes of Drowning of the Calcare Massiccio Platform in Northern Apennines," *Bollettino della Societa Geologica Italiana* 124 (2005): 247–58.
31. L. Carmignani and R. Kligfield, "Crustal Extensions in the Northern Apennines: The Transition from Compression to Extension in the Alpi Apuane Core Complex, *Tectonics* 9, no. 6 (1990): 1275–303.
32. Carlo Baroni, Giuseppe Bruschi, and Adriano Ribolini, "Human-Induced Hazardous Debris Flows in Carrara Marble Basins (Tuscany, Italy)," *Earth Surface Processes and Landforms* 25 (2000): 93–103.
33. Other marbles may have impurities of iron or manganese oxides, which generate red, yellow, cream, or pink marbles. So called "green marble" is the metamorphic rock serpentine.
34. John Logan, "On-Site and Laboratory Studies of Strength Loss in Marble on Building Exteriors," in *Fracture and Failure of Natural Building Stones: Applications in the Restoration of Ancient Monuments*, ed. S. K. Kourkoulis, (Berlin: Springer, 2006), 345–62.
35. Stone, *Evolution of an Architect*, 141.
36. Standard Oil's 1972 annual report lists 46,627 employees. No number was kept on how many had training in geology, but it seems logical that many of them must have had a geologic background. How else do you find oil?

## 9: READING, WRITING, AND ROOFING

1. Slate pencils were made of soapstone or softer slate and left a light, erasable mark on the tablets. Curiously, between 1867 and 1904, the *New York Times* reported on six children dying from slate pencil wounds; nearly all of the youth fell on their pencils by accident.
2. Stephen Ambrose, *Duty, Honor, Country: a History of West Point* (Baltimore: The Johns Hopkins Press, 1966), 19.
3. Another missionary, Reverand John Williams, wrote in his memoir that students on the Cook Islands in 1833 lacked chalk for writing so broke off the spines of a sea urchin, burnt them slightly, and used the spines for writing. The spines came from the slate pencil urchin, *Heterocentrotus*

*mammillatus*, and had been used for centuries as files to make bone and shell fishhooks.

4. Modern blackboards are no longer made of slate. They may be ceramic-coated steel, or enameled or painted composites. Manufacturers consider the ceramic boards to be more durable, with a shelf life of fifty years. Not bad unless you consider that slate boards from the early 1900s still look as good as the day they went up.
5. I could not have written about the geology of slate without several thorough and helpful phone conversations with Jack Epstein. Dr. Epstein also read a draft of this chapter.
6. The statistics on use of slate, as well as subsequent information on the change from Welsh to American slate, come from Jeffrey S. Levine, "An history of the United States slate industry (1734–1988)" (Master's thesis), Cornell University, 1988).
7. Boston Building Ordinances, 1631–1714, *Journal of Society of Architectural Historians* 20, no. 2 (1961): 90–91.
8. Samuel Carpenter built the house, which later served as home for William Penn and William Trent, founder of Trenton. It was razed in 1867.
9. There is little evidence that the Peach Bottom Slate won any award in London in 1851. According to the official archives of the Crystal Palace Exhibition, no slate from Pennsylvania was entered at the show. The earliest reference to the award that I could find is an article in the July 30, 1910, magazine *The Mining World*. Seems a bit odd that there is no mention of this from any newspaper of the day. I suspect we will never know exactly how the story got started.
10. The red Indian River slate formed 464 million years ago. The purple and green slates come from a rock unit known as the Middle Granville and formed 530 million years ago. Vermont produces a black slate that was deposited 540 million years ago. Principal metamorphosis occurred 450 million years ago during the Taconic orogeny.
11. Two black slates come from Pennsylvania, the 445-million-year-old Martinsburg and the 455-million-year-old Peach Bottom Slate. Virginia's slate formed 450 million years ago and Maine's 30 million years later, with metamorphosis occurring about 370 million years ago during the Acadian orogeny.
12. In 1931 the British Standards Institution sought to establish a uniform system for roofing shingles. Abandoned were terms such as "lady," "countess," and "duchess." They had arisen in the 1700s replacing the Catch-22-esque "single," "double," "double doubles," and "double double doubles," as well as the equally poetic "farewell," "mope," "haghattee," and "Jenny-why-gettest-thou."
13. Ben Kantner, interview with author, Snohomish, Washington, January 2008.
14. Many people choose copper over stainless steel because they like the look, especially when using copper gutters. Like stainless steel, copper doesn't rust. It also inhibits moss growth.
15. When it became known that John Quincy Adams purchased a billiard table and placed it in the White House in 1825, representative Samuel Carson of

North Carolina condemned the president for a sin that "would shock and alarm the religious, the moral, and reflecting part of the community."

16. Gravestone researchers refer to documentation of slate coming from Slate Island as early as 1630. In addition, twenty tons of slate were quarried near Braintree and on Hangman's Island in 1721. These dates raise a question about the claim that the Peach Bottom quarry is the first in America. Which was first probably depends on how one defines quarry.
17. Walter Isaacson, *Benjamin Franklin: An American Life* (New York: Simon & Schuster, 2003), 470.
18. Written on July 26, 1838, by Enoch Cobb Wines and published in *A Trip to Boston in a Series of Letters to the Editor of the United States Gazette* (Boston: Little and Brown, 1838), 45.
19. Blanche Linden-Ward, *Silent City on a Hill: Landscapes of Memory and Boston's Mount Auburn Cemetery* (Columbus, OH: Ohio State University Press, 1989), 218.
20. Mount Auburn Cemetery lifted its slate ban in the 1870s during a period of colonial revival resulting from the country's centennial celebrations. Other cemeteries also returned to using slate. The new slate grave markers preserved the shape of the old, but not the symbols of the Puritans. Floral decorations replaced death's-heads. Most of the slate came from eastern Massachusetts and Rhode Island, as well as Maine and Pennsylvania. Some carvers also imported slate from Wales.

## 10: "AUTUMN 20,000 YEARS AGO"

1. Chris Romanek, phone interview with author, April 2007.
2. Translations of Vitruvius are based on Ingrid D. Rowland, trans., *Vitruvius: Ten Books on Architecture* (New York: Cambridge University Press, 1999) or from the various papers of Marie Jackson.
3. I could not have written this section on travertine, Vitruvius, and Roman building techniques without the support of Dr. Jackson. She graciously answered numerous questions and read the text and corrected my interpretations. Key works include: M. D. Jackson, F. Marra, R. L. Hay, C. Cawood, and E. M. Winkler, "The Judicious Selection and Preservation of Tuff and Travertine," *Building Stone in Ancient Rome, Archaeometry* 47, no. 3 (2005): 485–510; Marie Jackson and Fabrizio Marra, "Roman Stone Masonry: Volcanic Foundations of the Ancient City," *American Journal of Archaeology* 110 (2006): 403–436; Marie Jackson, Cynthia Kosso, Fabrizio Marra, and Richard Hay, "Geological Basis of Vitruvius' Empirical Observations of Material Characteristics of Rock Utilized in Roman Masonry," *Proceedings of the Second International Congress of Construction History* 2 (2006): 1685–1702.
4. Janet DeLaine, "The Supply of Building Materials to the City of Rome," in *Settlement and Economy in Italy 1500 B.C. to A.D. 1500*, ed. N. Christie. Papers of the Fifth Conference in Italian Archaeology, *Oxbow Monograph* 41 (1995): 554–62.

5. Information on cutting and use of dowels is from Lynne Lancaster, "The Process of Building the Colosseum," *Journal of Roman Archaeology* 18 (2005): 57–83.
6. Robert Folk, phone interviews with author, March 2007.
7. Henry S. Chafetz and Robert L. Folk, "Travertines: Depositional Morphology and the Bacterially Constructed Constituents," *Journal of Sedimentary Petrology* 54 (1984), 289–316.
8. In a notorious case in the mid-1990s, which mimics what happened in Chicago to Big Stan, a high-rise office building in Boston had to have every travertine panel reanchored because of frost-induced cracking. Some panels required extra bolts because of numerous cracks and other panels had to be replaced because of excess damage.
9. Dante mentions the Bulicame in Canto XIV of the *Inferno*. He wrote of a brook visited by prostitutes, and more appropriately, "Its bed and both its banks were made of stone, together with the slopes along its shores." That stone, of course, was travertine.
10. Paula Noble, phone interview with author, August 2007.
11. Robert L. Folk, "SEM Imaging of Bacteria and Nannobacteria in Carbonate Sediments and Rocks," *Journal of Sedimentary Petrology* 63 (1993), 990–99.
12. Brenda Miller, phone interview with author, May 2007.
13. Many publications state that the Getty travertine came from the same quarry as the Colosseum travertine. This is not technically correct. After two thousand years, no original quarries remain from the time of the Romans. Both the Colosseum and Getty stone came from the historic Barco region.
14. Richard Meier, *Building the Getty* (Los Angeles: University of California Press, 1999), 186.
15. Meier quote is from a brochure produced by the Getty Museum titled *Architecture of the Getty*.
16. Michael Palladino, interview with author, Los Angeles, March 14, 2007.
17. Quote comes from Meier's acceptance speech for the Pritzker Prize, which he won in 1984.
18. Meier, *Building the Getty*, 57.
19. The first big Italian travertine project was New York's Pennsylvania Station in 1910. Penn Station led to a minor boom in travertine imports but mostly in smaller projects, ironically including the interior of the county courthouse in Bedford, Indiana, built in 1930.
20. Fabrizio Mariotti, interview with author, Tivoli, Italy, October 18, 2007.
21. Mariotti is correct that deposition occurred twenty thousand years ago, but the primary period of deposition appears to be older. Geologists have not focused on dating specific layers of travertine. Generally, deeper layers are older, though there is not a direct correlation between age and depth.
22. Meier, *Building the Getty*, 104.

# Index

Note: Page numbers in *italics* refer to photographs.